1st Semester Electronics

Mark L Gray

First published in United States of America February 2014

Copyright © Mark L. Gray February 2014

All rights reserved. No part of this publication may be reproduced, stored in a retrieval system or transmitted in any form or by any means without the prior permission in writing of the publisher, nor be circulated in writing of any publisher, nor be otherwise circulated in any form of binding or cover other than that in which it is published without a similar condition including this condition, being imposed on the `subsequent purchaser.`

Table of Contents

Pg. 3-**Introduction**

Pg. 4-Quick Study Points to Remember
Pg. 6-Lesson 1-Dealing with milli, micro, nano & pico
Pg. 7-Lesson 2-Electronic/Electrical Quantities
Pg. 8-Lesson 2a-Conventions Used in this Text
Pg. 8-Lesson 3-Recognizing Circuit Types
Pg. 8-Lesson 3a-Recognizing Circuit Types-Series Circuits
Pg. 10-Lesson 3b-Recognizing Parallel Circuits
Pg. 11-Lesson 3c-Recognizing Circuit Types-Series Parallel Circuits
Pg. 13-Lesson 3d-Recognizing Circuit Types-Parallel Series Circuits
Pg. 14-Lesson 4-Ohm's Law
Pg. 19-Lesson 5-Analyzing Series DC Resistive Circuits
Pg. 20-Lesson 5a-Series Circuit with Total Current (I_T and Total Power (P_T) unknown
Pg. 21-Lesson 5b-Series Circuit with Unknown Wattages
Pg. 23-Lesson 5c-Series Circuit with Unknown Resistances
Pg. 24-Lesson 5d-Series Circuit with Unknown Source Voltage
Pg. 25-Lesson 5e-Series Circuit with Unknown Potentiometer Value
Pg. 26-Lesson 6-Simple Parallel Resistive Circuits
Pg. 27-Lesson 6a-Parallel Circuit Example #1
Pg. 29-Lesson 6b-Parallel Example #2 Parallel Lights Circuit
Pg. 30-Lesson 7-Series-Parallel Resistive Circuits
Pg. 33-Lesson 8-Parallel-Series Circuits
Pg. 35-Lesson 9-Power in Resistive Circuits
Pg. 36-Lesson 9a-P_T with I_T & V_T Known
Pg. 37-Lesson 9b-PT & PRX with RT & VT Known
Pg. 38-Lesson 9c-PT & PRX with RT & IT Known
Pg. 39-Lesson 9d-PT in Parallel Circuits
Pg. 40-Lesson 10-Series Resistive Voltage Divider Circuit
Pg. 42-Lesson 11-Series and Parallel DC Voltage Sources Amp Hour
Pg. 43-Lesson 11a-Amp Hour Ratings
Pg. 44-Lesson 12-Parallel Current Sources
Pg. 48-Appendix A-Series Circuit Analysis Unknown Quantities Flowchart
Pg. 49-Appendix B-Practice Series Circuits
Pg. 51-Appendix C-Parallel Practice Circuits
Pg. 54-Appendix D-Series Parallel Practice Circuits
Pg. 56-Appendix E-Parallel Series Circuits
Pg. 60-Appendix F-Practice Unit Conversions
Pg. 62-About the Author

Introduction

This text is intended to help 1st semester electricity and electronics students or anyone who has the need to learn to analyze basic DC resistive circuits quickly and efficiently. My goal is to present these analysis methods in an easy to follow format that I have proven work over years of presentation in the classroom. I know these "to the point" procedures to be time effective and I believe are much easier and practical for the beginner to learn than the methods presented in most textbooks.

What this text will not cover is the type of material that is presented as fill in the blanks, multiple choice or essay questions on exams. Anyone who puts out the effort can learn to pass those types of tests, but it is circuit analysis that is the make or break topic for many students.

You do not need an advanced calculator to work the problems in this text. A simple calculator such as the TI-30 is sufficient.

Every circuit variation has a proper procedure to analyze it, the only variations are the known values that are given to work with which will result in using different Ohm's Law formulas to solve.
The lessons presented start at the most basic resistive circuits and work toward the more complex series/parallel circuits. It is impossible to cover every variation in these circuits but if you learn the principles presented here you should be able to solve most with little difficulty after some practice.

The circuit drawings in this text were made using Multisim™, which is a product and trademark belonging to National Instruments Inc. I have used this program for many years in my Electronics classes and I believe that it is a tremendous asset to Electronics education. Nothing beats being able to draw a virtual circuit that actually runs and being able to use virtual instruments to analyze the circuit. I urge you to buy a student copy and take the time to learn to use it as Multisim™ will help you so much in your Electronics education.

Remember-this text is organized into a series of short lessons for the 1st semester electronics or electrical students covering basic circuit analysis and isn't intended for advanced students except as a refresher.

Quick Study Points to Remember for Each Circuit Type

Ohm's Law Points to Remember

• Ohm's Law is the basis for all circuit analysis
• You must have two known values for either the entire circuit or for an individual component to find the unknown value

Series Circuits Points to Remember
• There is only one current path and all components see the same current
• Voltage divides between the components
• The sum of the individual voltage drops equals the source voltage
• Total resistance, RT, is found by simply adding up the individual resistances
• The larger the resistance the larger the voltage drop across that resistance
• The sum of the individual powers equals the total power
• Higher RT equals lower total current and vice versa
• Voltage sources in series will add their individual voltages
• Any voltage sources that are in the circuit with its polarity reversed will subtract its voltage from the voltage sum

Parallel Circuits Points to Remember

• All branches see the same voltage
• Current in each branch depends on the resistance in that branch
• The total resistance, RT, is always less that the lowest resistance in the circuit
• Disconnecting any branch will cause RT to increase and IT to decrease
• Adding branches will cause RT to drop and IT to increase (as more current paths are added)
• A resistor having an open fault will remove that branch from the circuit
• Components are connected in parallel when their terminals share the same connection points
• Identical voltage sources in parallel have a VS equal to one of the source V values. Voltage sources in parallel are usually the same value.
• All the above is true no matter how many branches there are

Series Parallel Circuits Points to Remember

• Series-Parallel circuits are Series circuits with one or more parallel sections.
• The parallel portion(s) of the circuit must be resolved first
• RT is the sum of the equivalent parallel resistance and all series resistances
• The series resistance(s) see IT
• The parallel portion(s) see IT which is divided proportionally in each branch based on each resistive value
• This is a Series-Parallel circuit because the parallel portion is in series with the series resistance(s)

Parallel Series Circuits Points to Remember

• Each branch sees the source voltage
• Reduce each series branch to one equivalent resistance by adding the resistances
• The voltage (Vs) on each branch is divided between the series components based on their values just as a standard series circuit
• Once each branch is reduced to a single resistance solve for RT just like a parallel circuit.
• RT will always be less than the RT of the branch with the least resistance just as with a standard parallel circuit

Circuit Power Points to Remember

• Use the P (watts) section of the Ohm's Law wheel to calculate power
• Power can be calculated for an entire circuit or an individual resistance
• The sum of the individual resistive powers will equal PT
• The largest resistance will dissipate (use) the most power
• Power in purely resistive circuit is True power
• Power is constant in DC circuits with an unchanging RT or Vs
• The more True power that is used in a circuit the more heat that will be generated
• The physical size of a resistor determines how much wattage it can handle
• Wattage rating of a resistance has no bearing on basic circuit analysis

Lesson 1 Dealing with milli, micro, nano & pico

In studying electricity and electronics you will constantly be dealing with various units and quantities and getting comfortable with doing so is an absolute necessity. We will stay with just the raw numbers here as, depending on the instructor, you may or may not use scientific or engineering notation. You can properly learn the use of either in a technical math class, which is why all Engineering curriculums require at least one math class. Whatever method you are required to use, understanding the below number range in a must.

Greater Than 1 | Less Than 1

000.000.000.000.000.000.000

Mega	Kilo	Deca	milli	micro	nano	pico
M	K	(rarely used)	m	u	n	p

Upper Case Greater Than 1 | Lower Case Less Than 1

The chart above illustrates the units and their positions both greater than and less than one. You should get used to using the proper abbreviation ASAP as writing out the entire numeric value gets cumbersome and is not very professional but is initially far easier to learn for most people.

Now let's work through some examples starting from greater than one using volts as the unit:

10,000,000 Volts = 10 million volts. The proper expression is 10MV.
1,000,000 Volts = 1 million volts. The proper expression is 1MV.
100,000 Volts = 100KV
10,000 Volts = 10KV
1,000 Volts = 1KV
100 Volts would be expressed as 100V. Values of less than 1000 are usually expressed as is and the term Deca is not commonly used.

Now here is where most start having problems when the value is less than 1.

Examples:
.100V equals 100 millivolts from looking at the chart above. The preferred expression is 100mV. Notice that the m is lower case which is VERY important. The lower case m stands for milli and

the upper case M stands for Mega (million). Expressing the value as .100V is technically correct but not accepted as the preferred expression in the Engineering community. Again, you still need to understand the raw numbers.

.00001 should be expressed as 10uV and not .01mV
.000001V = 1uV
.0000001V = 100nV
.00000001V = 10nV
.000000001V = 1nV
.0000000001V = 100pV
.00000000001V = 10pV
.000000000001 = 1pV

You can see how writing out the entire digit string gets unwieldy in a hurry so you must learn to abbreviate properly as soon as you can. Most instructors will require you to anyway.

As you are learning use the conversion chart at the beginning of this lesson to help you determine the numeric positions and the proper abbreviation.

See Appendix F for practice conversions.

Lesson 2 Electrical/Electronic Quantities

Amperage is the flow of electrical current (electrons) and is the amount of electrons that are flowing. Abbreviated as A. Remember that uppercase I represents current and the unit is A (short for amps).

Resistance is literally the resistance to the flow of electrical current. The unit is ohms the symbol used to represent resistance is the Greek letter omega ().

Voltage is electrical pressure or potential and is represented by the upper case letter V. Older textbooks used E for Energy instead of V.

Power is the product of Voltage and Current and the unit is Watts. The upper case letter W represents Watts.

Lesson 2a Conventions Used in This Text

RT = Total resistance   VT = Total voltage   IT = Total current   PT = Total power   RX = any Resistor or resistance   VRX = Voltage on a given resistor x   PRX = Power dissipated by a given resistor

IRX = Current through a given resistor   Vs = Source (battery or power supply) voltage

I branch = current in a given circuit branch   V branch = Voltage on a given circuit branch

Schematic Symbols Used:

 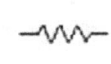 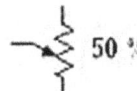

DC Power   Lamp   Resistor   Potentiometer   Ground

(Battery)   (Variable Resistor)

Lesson 3-Recognizing Circuit Types

Lesson 3a-Recognizing Circuit Types-Series Circuits

- Series circuits have only one current path from the voltage source to circuit ground-always!
- Circuit 1 is the most basic series circuit.
- Circuit 2 adds a light bulb-still a series circuit
- Circuit 3 has two voltage sources but still only one current path
- Circuit 4 add a potentiometer which is a variable resistor
- Circuit 5 adds a switch to switch between two resistive branches. No matter what position the switch is in there in circuit 5 there is still only one current path.

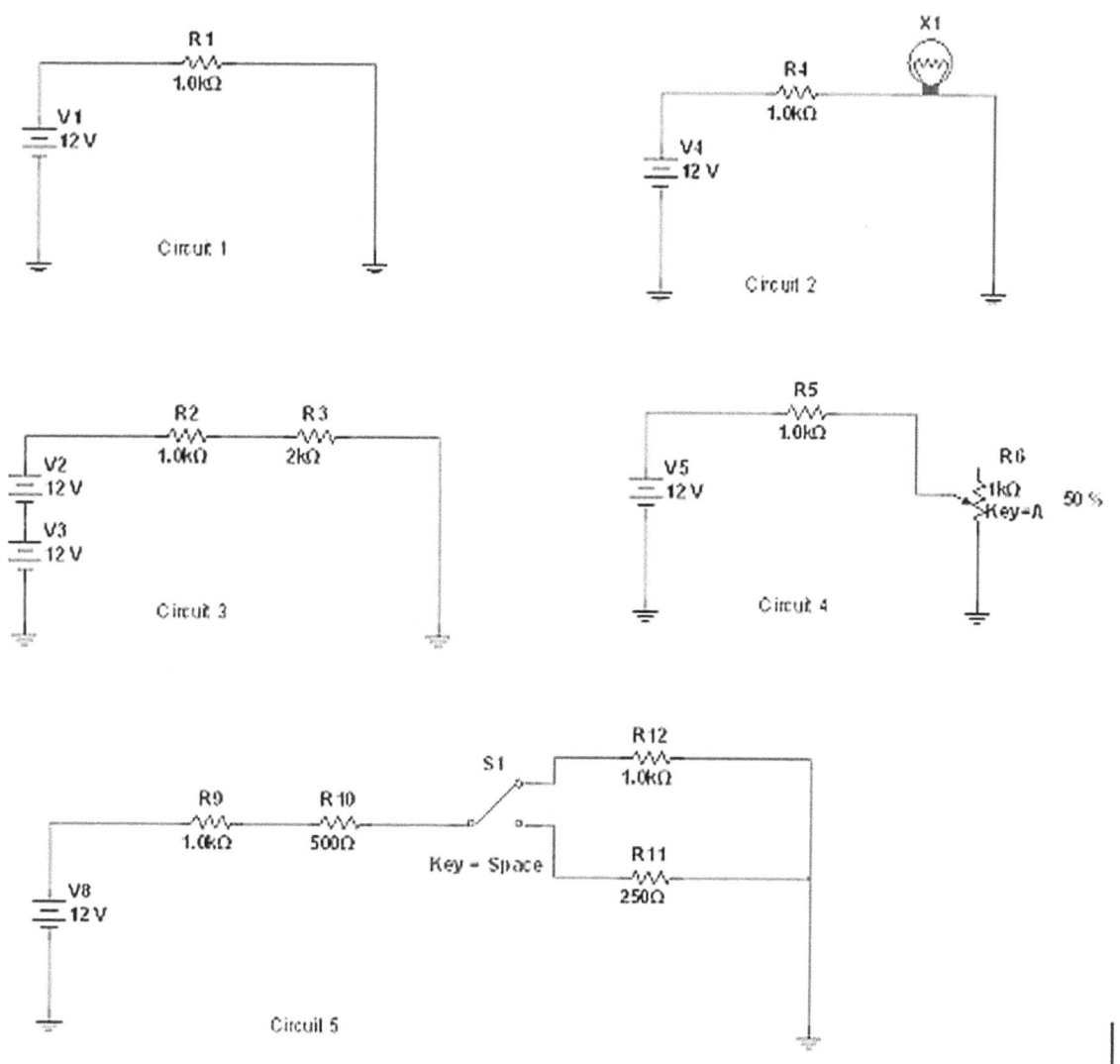

Shown on this page are various Series circuits configurations.

The common thread is that there is only one current path in each circuit from the DC source to ground.

Lesson 3b-Recognizing Parallel Circuits

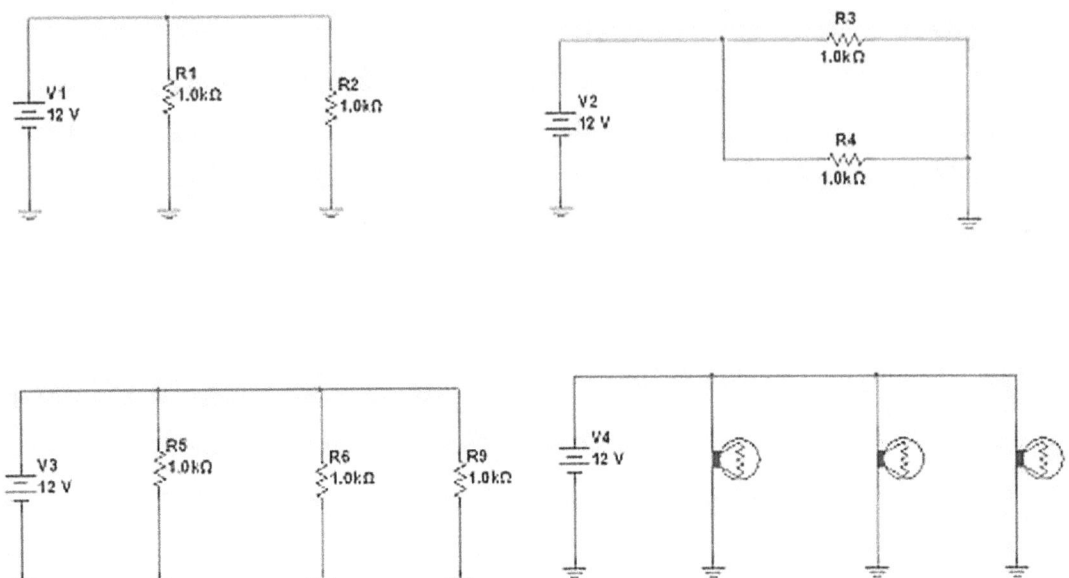

Parallel Circuits Points to Remember:

• Each branch in a Parallel circuit share a common connection to the voltage source and to the common ground point
• The current value in each branch depends on each branch resistance
• Each branch "sees" the same voltage
• Ground connections are common to all branches
• All the above is true no matter how many branches there are

Lesson 3c-Recognizing Circuit Types-Series Parallel Circuits

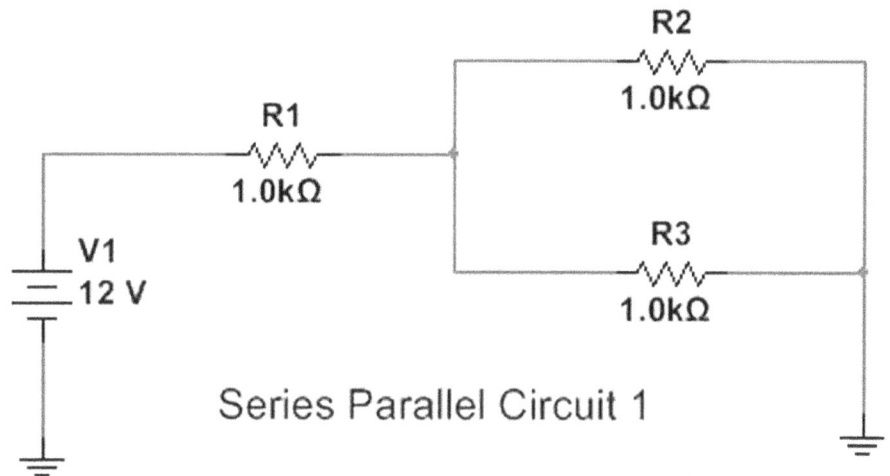

Series Parallel Circuit 1

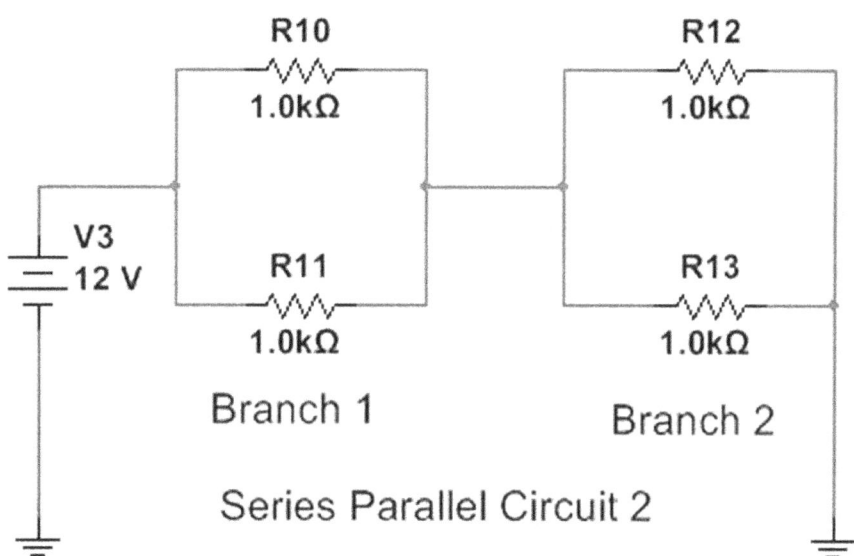

Series Parallel Circuit 2

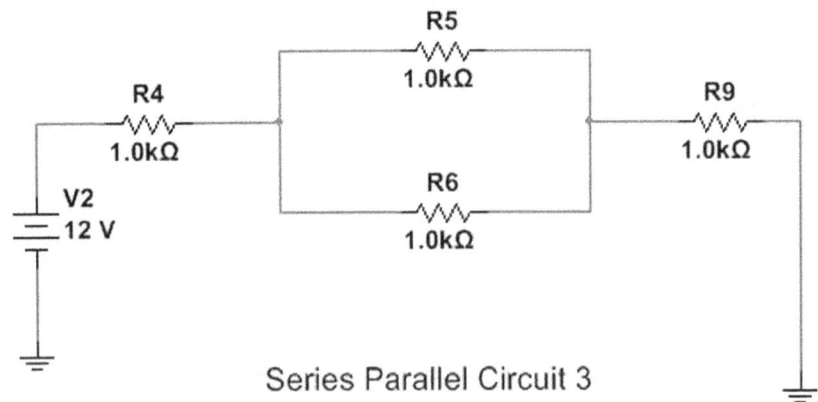

Series Parallel Circuit 3

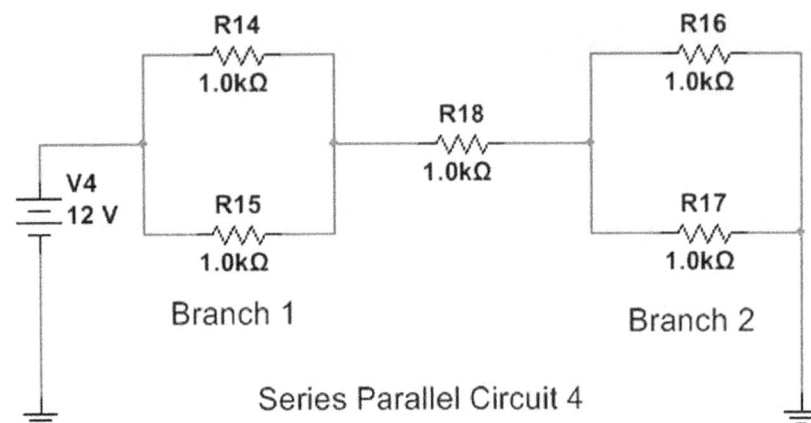

Series Parallel Circuit 4

Series Parallel Circuits Points to Remember:

- Series-Parallel circuits are Series circuits with one or more parallel sections.
- Circuit 1 has a series resistor R1 that sees the total circuit current pass through it. The R2/R3 parallel section is in series with R1.
- Circuit 2 has a parallel section, R5/R6, that is in series with R4 and R9.
- Circuit 3 has parallel Branch 1 in series with parallel Branch 2. The same total circuit current flows through each parallel branch which you will see later in the circuit analysis section.
- Circuit 4 has parallel Branches 1 & 2 in series with R18. The parallel sections are in series with R18 which makes this a Series circuit

Lesson 3d-Recognizing Circuit Types-Parallel Series Circuits

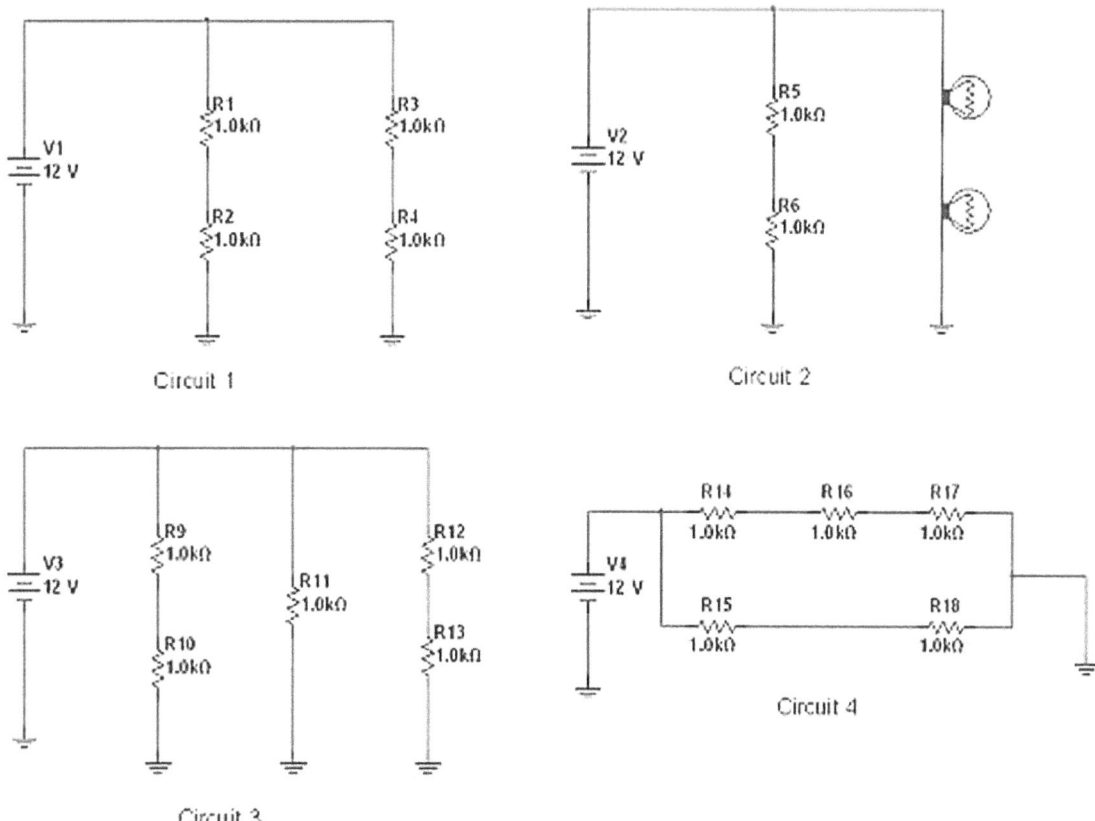

Parallel-Series Circuits Points to Remember

- Parallel-Series circuits are Parallel circuits with two or more series components in one or more branches.
- In Circuit 1 the branches are in parallel with components in series in each branch.
- Circuit 2 is the same configuration but with lamps in a branch
- Circuit 3 has three branches, 2 branches have series   components
- Circuit 4 has 2 parallel branches with series resistors in each branch

Lesson 4 Ohm's Law

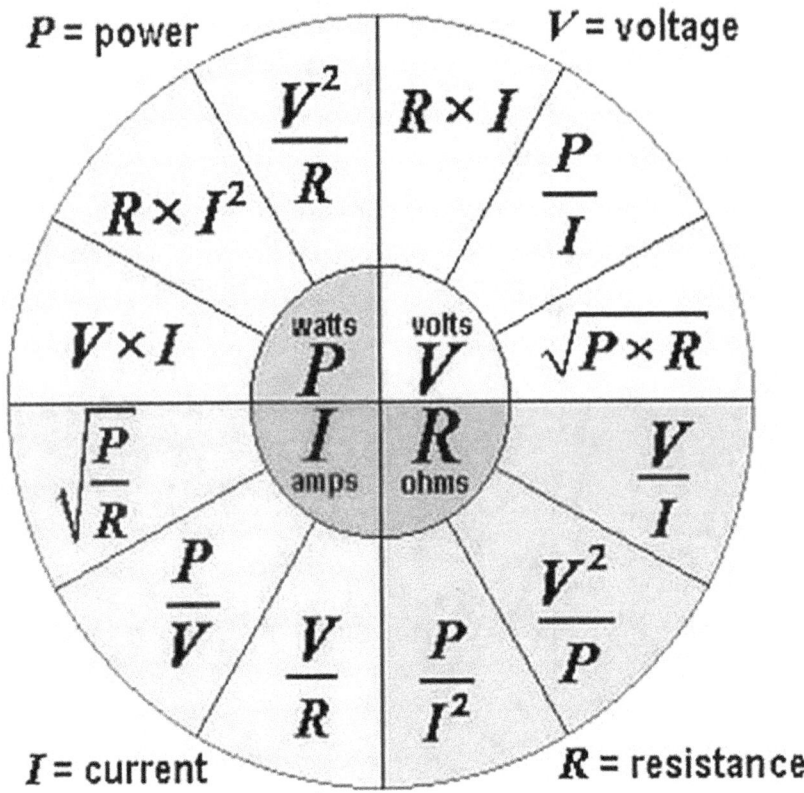

Ohm's Law Points to Remember:

• Ohm's Law is the basis for all circuit analysis
• You must have two known values for either the entire circuit or for an individual component to find the unknown value
• The center quantities in the wheel are the values being sought. Each has 3 possible ways to calculate using 2 known quantities.

All the variations in Ohm's Law shown in the wheel above are useful but in reality you can solve most calculations by remembering only 4 of the formulas:

V=I*R   R=V/I   I=V/R   P=V*I

You may have to add a few extra steps in a circuit analysis by only using these 4 formulas but if your instructor requires you to memorize Ohm's Law you can usually get by.

Now let's try some simple practice calculations.

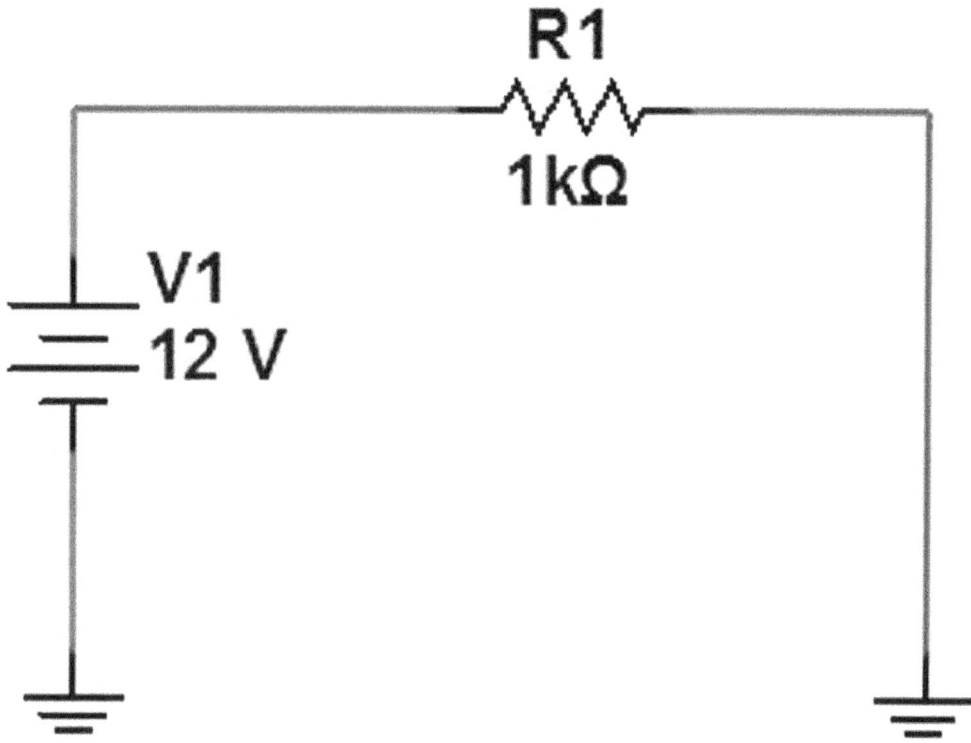

Example 1

In this example the two known values are voltage and resistance and we will be looking for the two unknown values of current and power.

Looking in the V section of the Ohm's Law chart we find that we need I=V/R and in the P section we need P=V*I.

I=12V/1KΩ=.012A or 12mA, P = 12V*12mA=.144W or 144mW

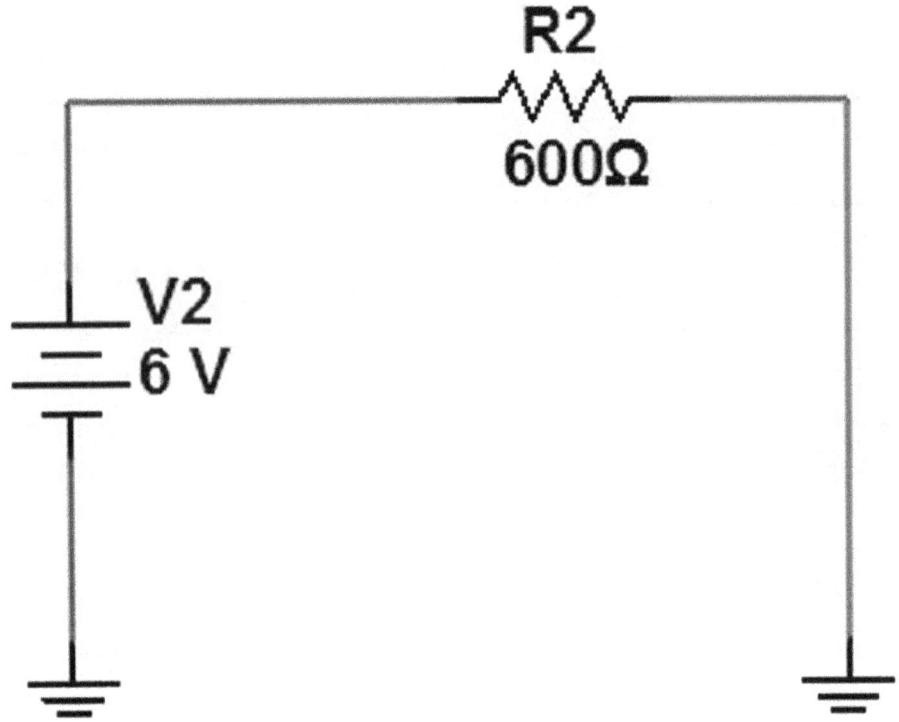

Example 2

Again looking in the V section of the Ohm's Law chart we find that we need I=V/R and in the P section we need P=V*I since I (current) and P (power) are the unknowns.

I=6V/600Ω = .010A or 10mA   P = 6V*10mA = .060W or 60mW

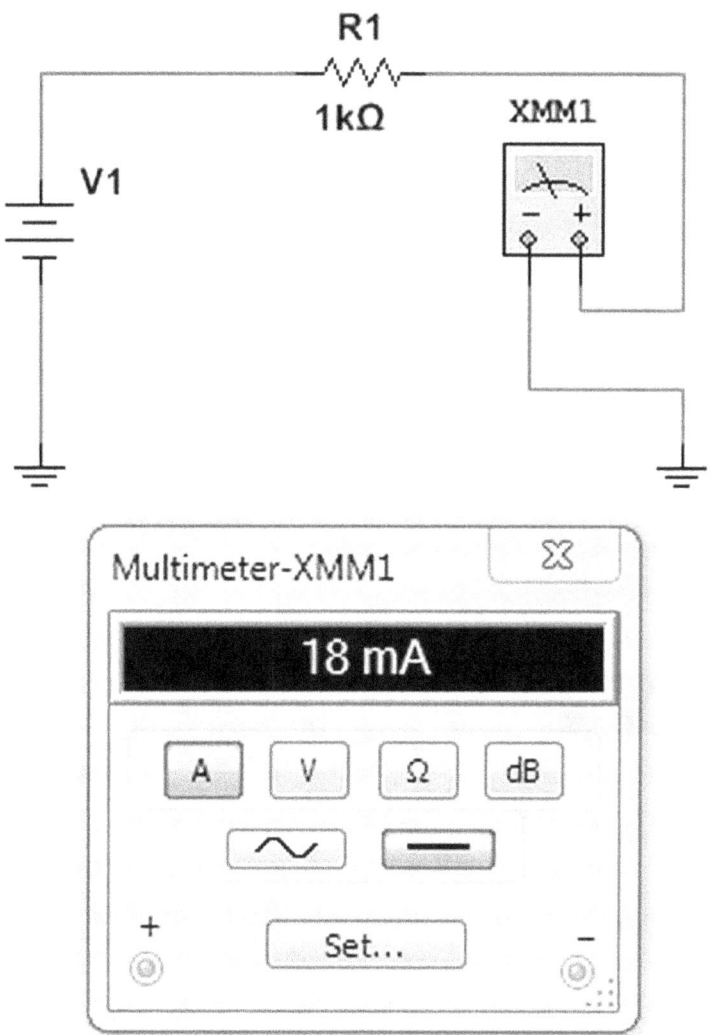

Example 3

Here we have an unknown DC voltage source value.

Our two knowns are resistance and current. Looking at the Ohm's Law wheel we find that V=I*R. 18mA*1KΩ = 18V.

The power in the circuit 18V*18mA = 324 mW or .324 Watts

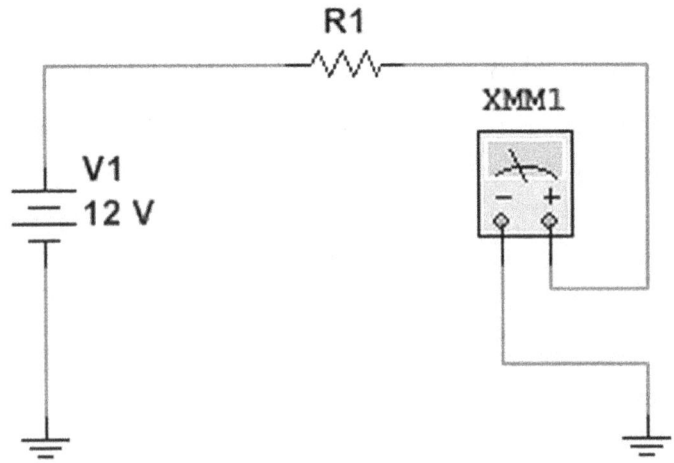

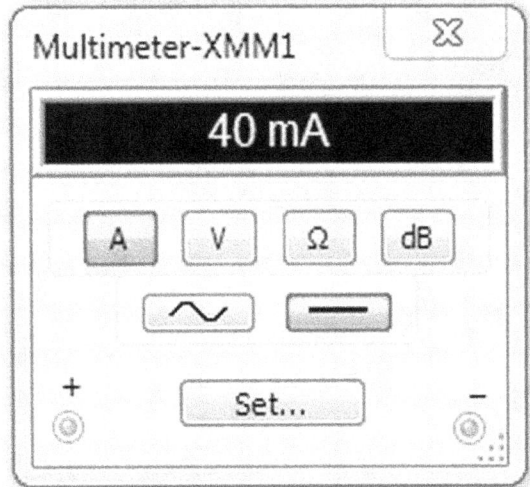

Example 4

The resistor value is unknown.

From the Ohm's Law wheel we get R=V/I. So 12V/40mA=300Ω. The power in the circuit is 12V*40mA = 480mW or .48 Watts

Remember-you must have two known values to get an unknown.

Lesson 5-Analyzing Series DC Resistive Circuits

Series Circuits Points to Remember:

• There is only one current path and all components see the same current
• Voltage divides between the components
• The sum of the individual voltage drops equals the source voltage
• Total resistance, RT, is found by simply adding up the individual resistances
• The larger the resistance the larger the voltage drop across that resistance
• The sum of the individual powers equals the total power
• Any OPEN fault will have the source voltage across the fault. Current flow will go to zero
• Any SHORT CIRCUIT fault across a component will cause the resistance of that component to go to near zero and RT to decrease. Total current will increase
• Higher RT equals lower total current and vice versa
• Voltage sources in series will add their individual voltages
• Any voltage sources that are in the circuit with its polarity reversed will subtract its voltage from the voltage sum

Remember:
One current path from voltage source to circuit ground in a Series circuit!

Lesson 5a-Series Circuit with Total Current (IT) and Total Power(PT)unknown

The first circuit usually taught in any electronics curriculum is the series circuit.

Formulas:
RT = R1 + R2 + R3 = RTIT = Vs/RT   VRX = IT*Rx
PT = IT*Vs

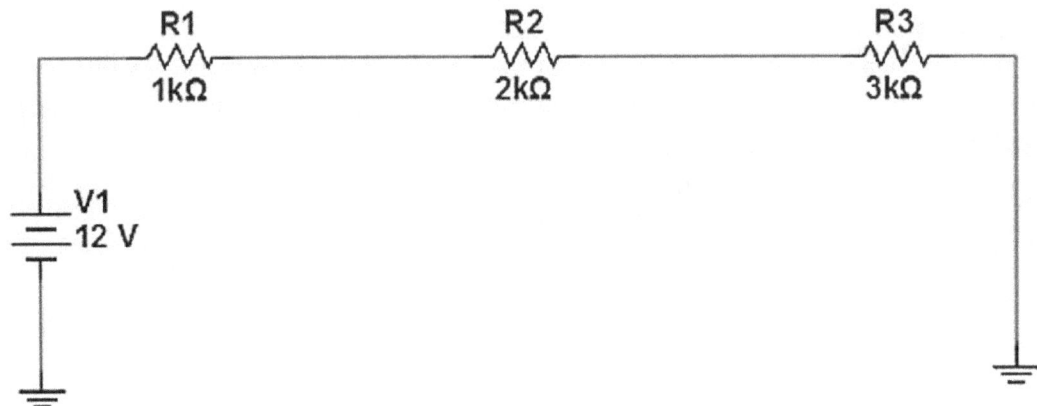

The recommended procedure to analyze a series circuit using the above series circuit as an example with Voltage and Resistance being the two known quantities:

1. Determine RT by adding the resistances. R1+R2+R3=RT. RT is needed to calculate IT.
2. Calculate IT, total current, by using IT = Vs/RT.
3. Calculate the individual voltage drops. VR1 = IT*R1,
VR2 = IT*R2 and VR3 = IT*R3
4. Check your calculations. Vs = VR1 + VR2 + VR3.
5. Calculate the power dissipated with each resistor by using the formula PRx = VRx*IT where x is the resistor being analyzed.
6. Check your calculations by adding the individual powers. Determine PT by using PT = Vs*IT. The two values should be the same. If they are the same then all of your other calculations are correct. Also, VR1+VR2+VR3 = Vs is another math check.

Answers: 1. RT = 6KΩ, 2.IT = 2mA, 3. VR1 = 2V, 4. VR2 = 4V, 5. VR3 = 6V 6. PR1 = 4mW, 7. PR2 = 8mW, 8. PR3 = 12mW

Use the procedure above for any series resistive circuit with known voltage and resistance values, the only variable is the number of resistors.
See Appendix C for practice series circuits.

Lesson 5b-Series Circuit with Unknown Wattages

This circuit type is analyzed differently as resistance and `current values are unknown` in this example. Remember that you must have two known values to determine an unknown quantity.

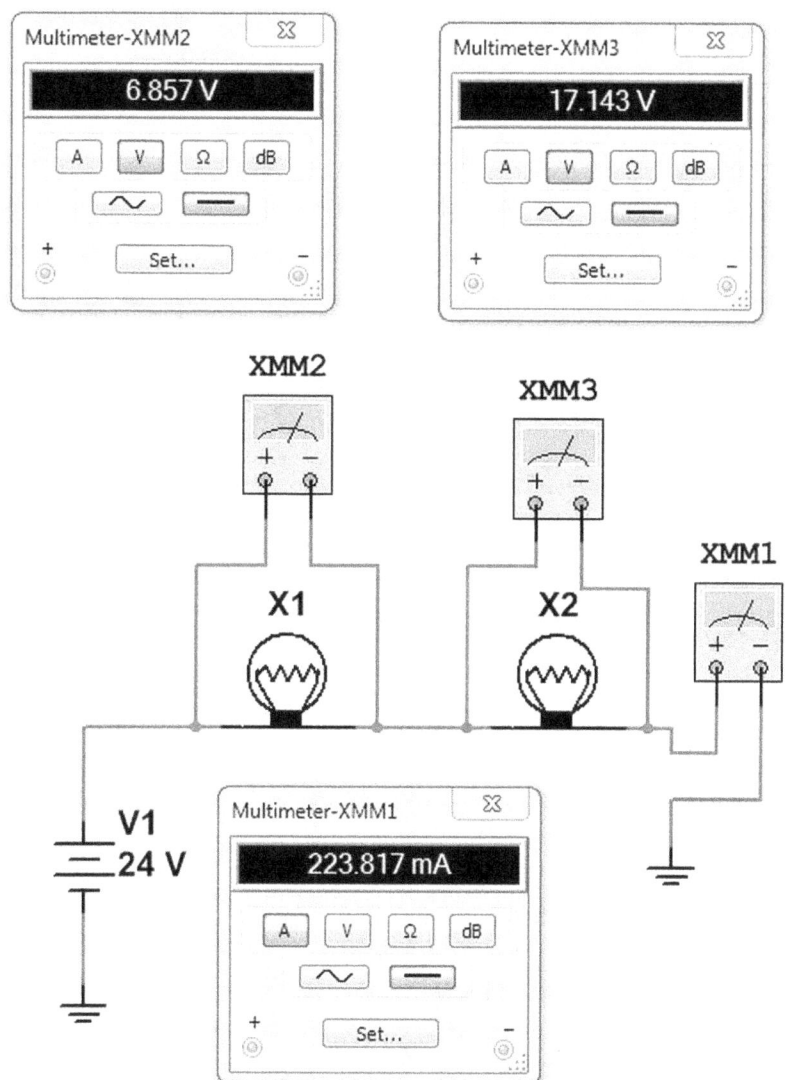

In the above circuit example we have the voltage measurement across each light and the total circuit current IT for our two known quantities. We want to solve for the individual and total circuit resistances and wattages.

Our two knowns are the source voltage and total current. The Ohm's Law formulas we will use are:

$P = V*I \qquad R = V/I$

The measured total current (IT) is 223.8mA

For L_1: $PL_1 = VL_1 * I_T$
PL1 = 6.86V*223.8mA = 1.53W

For L_2: $PL_2 = VL_1 * IT$
PL2 = 17.14V*223.8mA = 3.84W

$RL1 = V_{L1}/I_T$
RL1 = 6.86V/223.8mA = 30.65Ω

$RL2 = V_{L2}/I_T$
RL1 = 17.14V/223.8mA = 76.6Ω

We measured VR1 at 6.9V (rounded up). The following is the exact process to analyze.

1. $R_T = V_S/I_T$ which equals 24V/223.8mA = 107.24Ω

2. $R_{L1} = V_{L1}/I_T$ which equals 6.9V/223.8A = 30.8Ω

3. $R_{L2} = V_{L2}/I_T$ which equals 17.14V/223.8A = 76.6Ω

4. RT = RL1 + RL2 which equals 107.4Ω

5. Check your calculations - IT = VT/RT = 223.4mA which very closely matches the ammeter measurement. Rounding errors are always present.

6. PL1 = VL1*IT which equals 6.9V*223.8A = 1.54W

7. PL2 = VL2*IT which equals 17.14V*223.8mA = 3.84W

8. PT = Vs*IT which equals 24V*223.8mA = 5.37W

9. PT = PL1 + PL2 = 5.38W

Lesson 5c-Series Circuit with Unknown Resistances

Sometimes the color bands on resistors cannot be read for whatever reason. In this case we must measure the voltage drops on each resistor and measure the total current to determine the value of each resistor. Knowing the value of the voltage source isn't critical as the sum of the voltage drops will equal the source voltage.

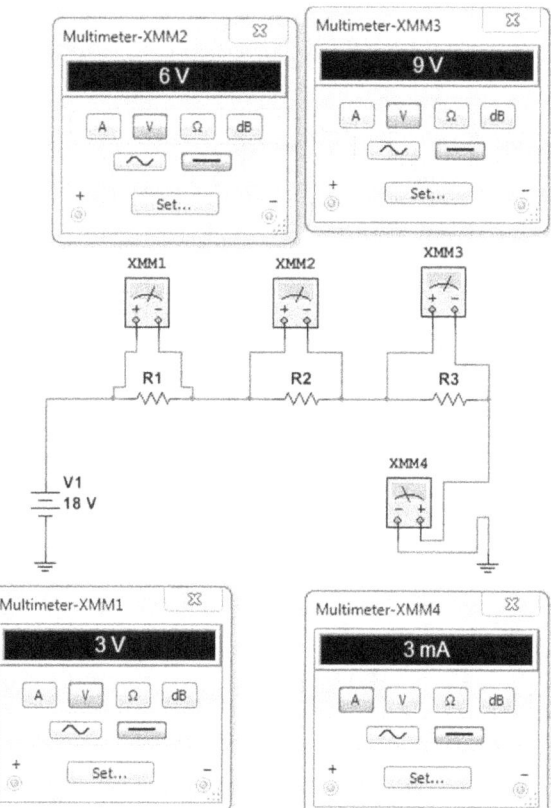

The Ohm's Law formulas we will use are:
P=V*I   R=V/I
The total current flowing through each resistor results in a voltage drop on each resistor based on each individual resistance.
1. R1 = VR1/IT which equals 3V/3mA(.003A). This gives a resistance value for R1 of 1KΩ(1000)

2. R2 = VR2/IT which equals 6V/3mA (.003A). This gives a resistance value for R2 of 2KΩ(2000Ω)

3. R3 = VR3/IT which equals 9V/3mA(.003A). This gives a resistance value for R3 of 3KΩ(3000Ω)

4. RT = R1 + R2 + R3 which equals 1KΩ+2KΩ+3KΩ=6KΩ

5. Vs = VR1 + VR2 + VR3 which equals 3V+6V+9V=18V

Check your math-IT = Vs/RT which equals 18V/6KΩ =.003A (3mA)

Lesson 5d-Series Circuit with Unknown Source Voltage

The unknown source voltage can be easily determined in this circuit by using the two known values of RT & IT (Ohm's Law V = I x R)

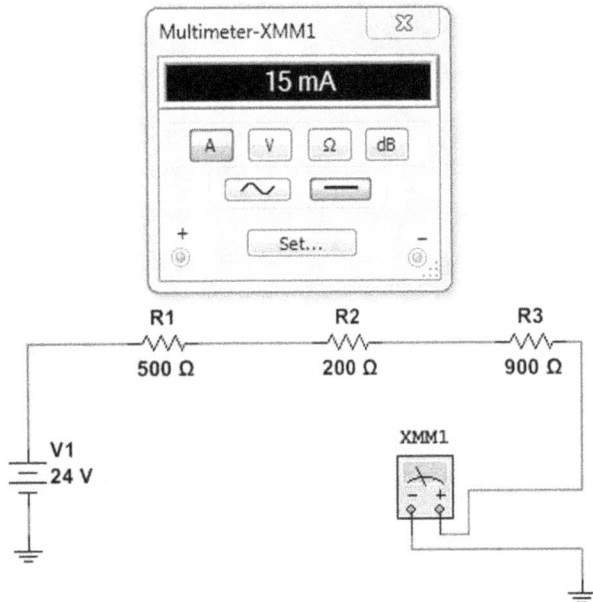

1. Determine RT, RT = R1 + R2 + R3 which equals 1600Ω or 1.6KΩ

2. Vs = IT x RT which equals 15mA (.015A) * 1.6KΩ = 24V

3. PT = Vs x IT which equals 24V x 15mA = .36W (360mW)

4. PR1 = IT² (15mA²)* R1(500Ω) = 112.5mW (.1125W)

5. PR2 = IT² (15mA²)* R2(200Ω) = 45mW (.045W)

6. PR3 = IT² (15mA²)* R3(900Ω) = 202.5mW (.2025W)

7. Check your math-calculated PT should equal PR1 + PR2 + PR3 which equals .1125W + .045W + .2025W = .36W

Lesson 5e-Series Circuit with Unknown Potentiometer Value

Potentiometers are variable resistors that are widely used to give current control in circuits. Most low price potentiometers have no easy way to determine what resistive values they have been set to as they don't have a reference scale printed on them. Here is an easy way to find out.

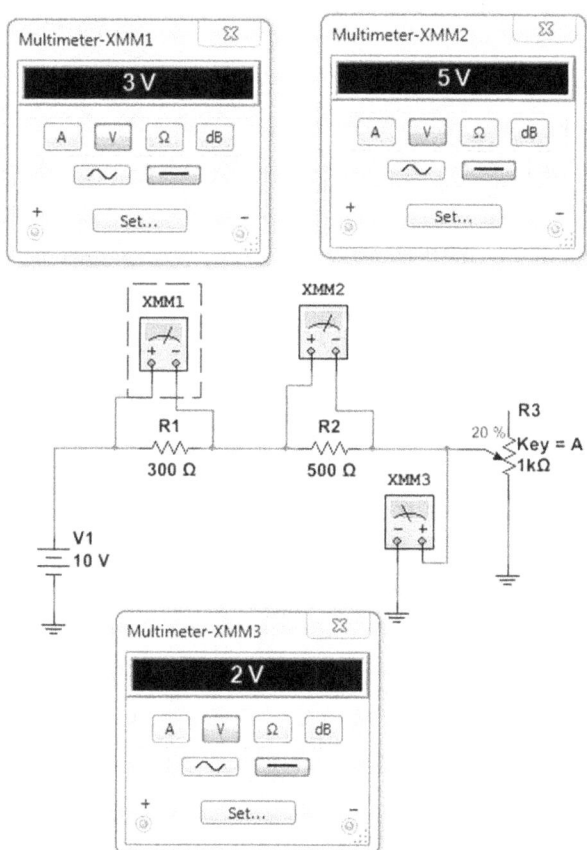

Knowing that current is the common element is any series circuit we will determine the value of

potentiometer R3 by using the voltage on R3 divided by the total current.

IT can be determined by using either VR1/R1 or VR2/R2.
IT = 3V/300Ω = 10mA or 5V/500Ω = 10mA

The resistance of the potentiometer R3 can be determined by the voltage across it.
R3 = 2V/10mA (.01A) = 200Ω

If you don't the potentiometer voltage but do know any of the resistor voltages here is another method:
If you know the voltage on R1: IT = 3V / 300Ω = 10mA.

For the given IT value of 10mA there is be a certain RT value to limit the current to this value.

Using Ohm's Law with the two knowns of 10V and 10mA to get the unknown RT (RT = VT/IT):

RT = 10V/10mA (.01A) = 1000Ω.

We know we have 800Ω in the total of R1 & R2 so the potentiometer R3 value will be 1000Ω - 800Ω = 200Ω.

Lesson 6-Simple Parallel Resistive Circuits

Parallel Circuits Points to Remember

- All branches see the same voltage
- Current in each branch depends on the resistance in that branch
- The total resistance, RT, is always less that the lowest resistance in the circuit
- Disconnecting any branch will cause RT to increase and IT to decrease
- Adding branches will cause RT to drop and IT to increase (as more current paths are added)
- A resistor having an open fault will remove that branch from the circuit
- Components are connected in parallel when their terminals share the same connection points
- Identical voltage sources in parallel have a VS equal to one of the source V values. Voltage sources in parallel are usually the same value.

Lesson 6a-Parallel Circuit Example #1

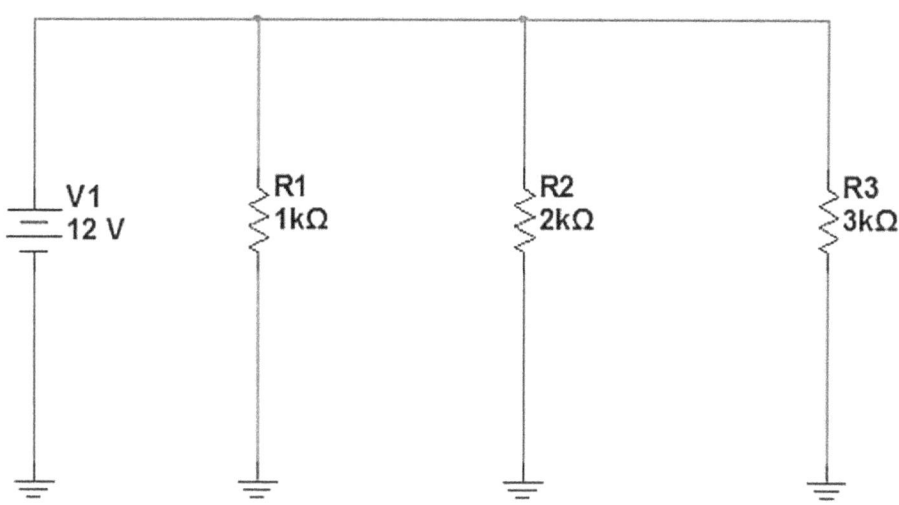

Formulas needed in this example:
IRx = Vs/Rx, RT = 1/R1 + 1/R2 + 1/R3 = 1/Rx, IT = Vs/RT or IT = IR1+IR2+IR3, PRx = IRx*Vs, PT = Vs*IT or PT = PR1+PR2+PRS, IT = Vs/RT

The recommended procedure to analyze a parallel circuit using the above example:
1. Note that all branches see Vs
2. Calculate RT, RT = 1/R1 + 1/R2 + 1/R3 = 1/Rx
3. Calculate IT, IT = Vs/RT
4. Calculate IR1, IR1 = Vs/R1
5. Calculate IR2, IR2 = Vs/R2
6. Calculate IR3, IR3 = Vs/R3
7. Check your calculations. Add IR1+IR2+IR3. This total should equal the IT calculated in step 3
8. Calculate total power using PT=Vs*IT
9. Calculate PR1
10. Calculate PR2
11. Calculate PR3
12. Check your calculations by adding PR1+PR2+PR3. The sum should equal the total power calculated in step 8

Answers: 2. RT=545.45Ω, 3. IT=22mA, 4. IR1=12mA, 5. IR2=6mA, 6. IR3=4mA 7. IT=22mA, 8. 264mW, 9. 144mW, 10. 72mW 11. 48mW. 12. 264mW

The complete circuit current measurements using Multisim:

1st Semester Electronics

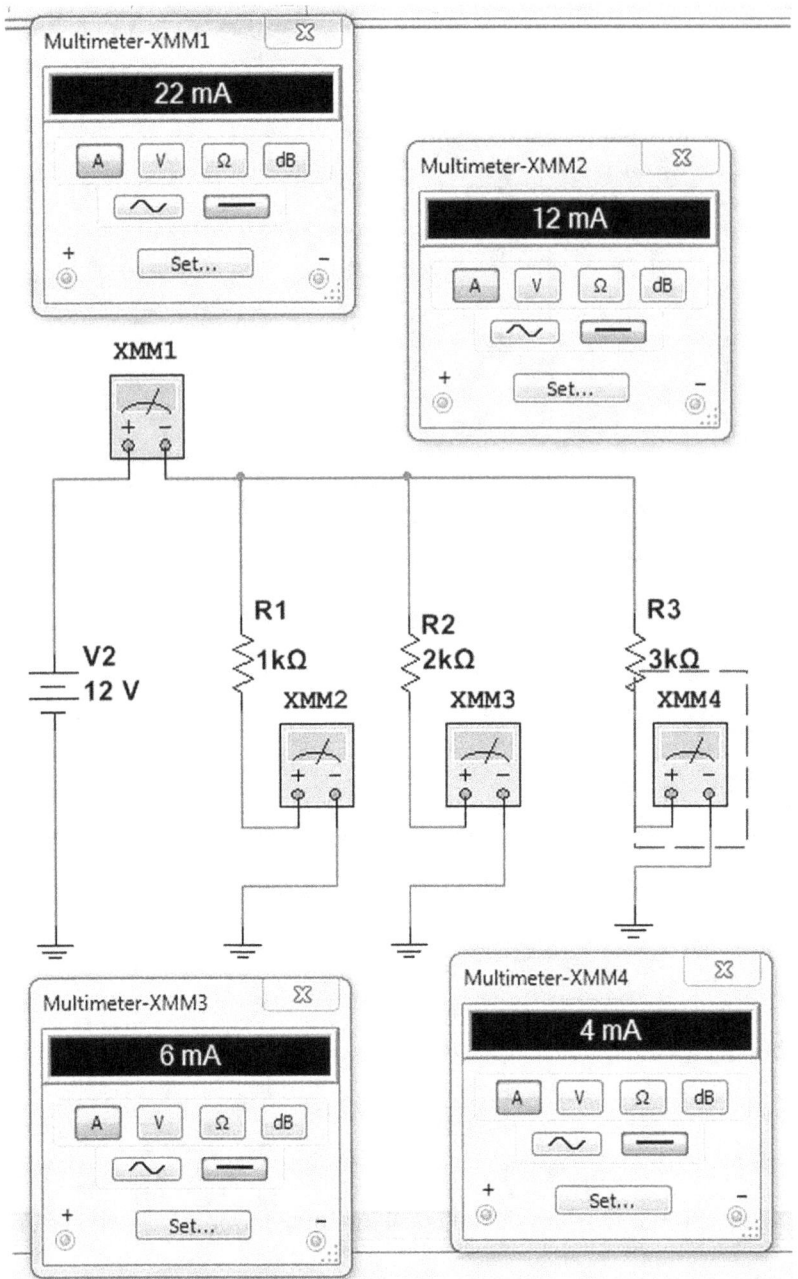

Lesson 6b-Parallel Lights Circuit Example #2

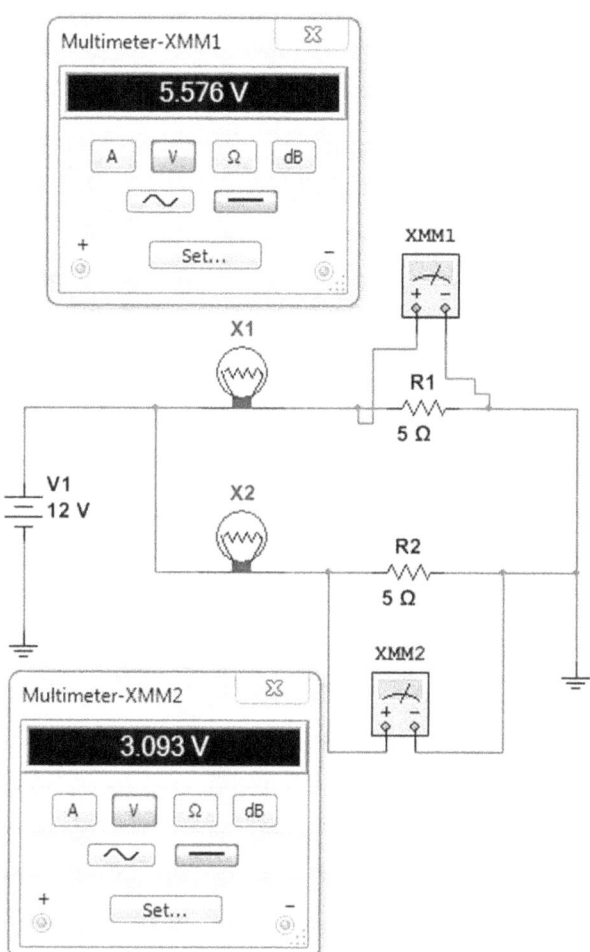

In this example we have two lights that are in separate branches that have the same voltage on each branch as is true with parallel circuits. We don't know the wattages of the lamps so we have inserted two sensing resistors to take voltage measurements to calculate each branch current. These resistor values are too small to affect the circuit parameters.
It is usually far more difficult to break open the branches to measure each current value, using sensing resistors is far more practical.

The formulas we will use are IBX = VRx/Rx for each branch current and PLX = VLX*IB for each lamp wattage.
1. IL1 = VR1/R1 which equals 5.576V/5Ω = 1.115A
2. VL1 = VS - VR1 which equals 12V - 5.576V = 6.42V
3. PL1 = VL1/IL1 which equals 6.42V*1.115A = 7.16W

4. IL2 = VR/R2 which equals 3.093V/5Ω = .6186A
5. VL2 = VS - VR2 which equals 12V - 3.093V = 8.91V
6. PL2 = VL2/IL2 which equals 8.91V *.6186A = 5.51W

See Appendix D for practice Parallel Circuits.

Lesson 7-Series-Parallel Resistive Circuits

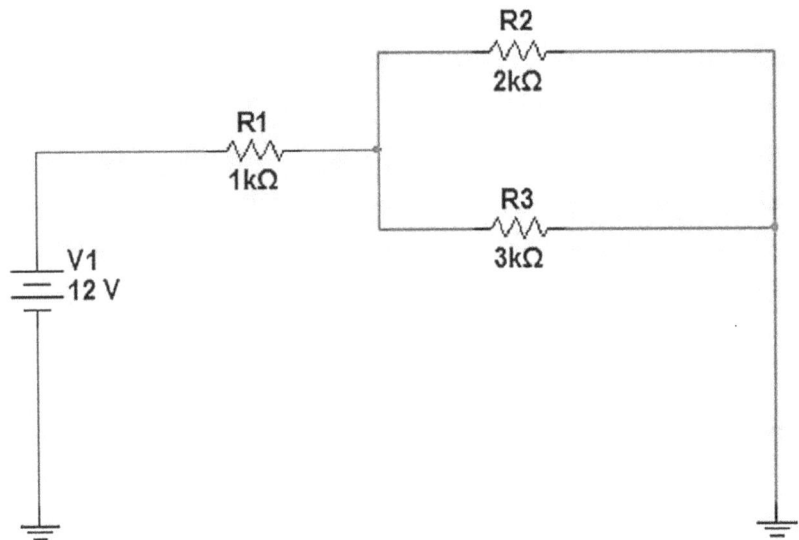

Series Parallel Circuits Points to Remember:

- The parallel portion(s) of the circuit must be resolved first
- RT is the sum of the equivalent parallel resistance and all series resistances
- The series resistance(s) see IT
- The parallel portion(s) see IT which is divided proportionally in each branch based on each resistive value
- This is a Series-Parallel circuit because the parallel portion is in series with the series resistance(s)

The recommended procedure to analyze a series-parallel circuit using the above example:

1. Solve RT for the parallel portion(s). 1/R2 + 1/R3 = Rparallel = 1.2KΩ
2. Add the parallel RT(s) to the series resistance(s) to get the circuit RT. R1 + Rparallel = 2.2KΩ
3. Redraw the circuit as a Series circuit (useful until you are competent with the analysis).

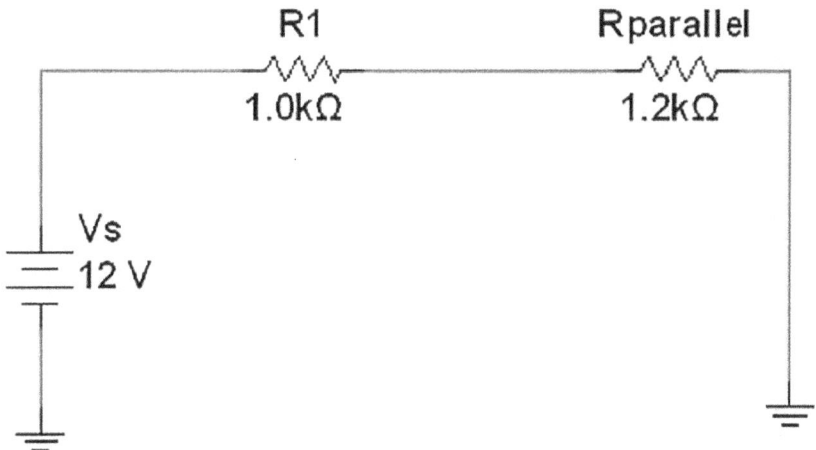

4. Calculate IT. IT = Vs/RT which equals .00545A or 5.45mA
5. Calculate the voltage drop for the series resistance
   VR1 = R1*IT = 1KΩ x 5.45mA = 5.45V
6. Calculate the voltage drop for the parallel portion
7. Vparallel = Rparallel*IT = 1.2KΩ*5.45mA = 6.55V
8. Check your math - The sum of the voltage drops should equal Vs.
9. Calculate the currents in each parallel portion. The sum of the branch currents in each parallel portion equals IT.
10. IR2 = Vparallel/R2 which equals 6.55V/2KΩ = 3.27mA
11. IR3 = Vparallel/R3 which equals 6.55V/3KΩ = 2.18mA
12. Check your math. The sum of the voltage drops should equal Vs. IR2 + IR3 + IT which equals 5.45mA.

The complete circuit measurements on the next page using Multisim:

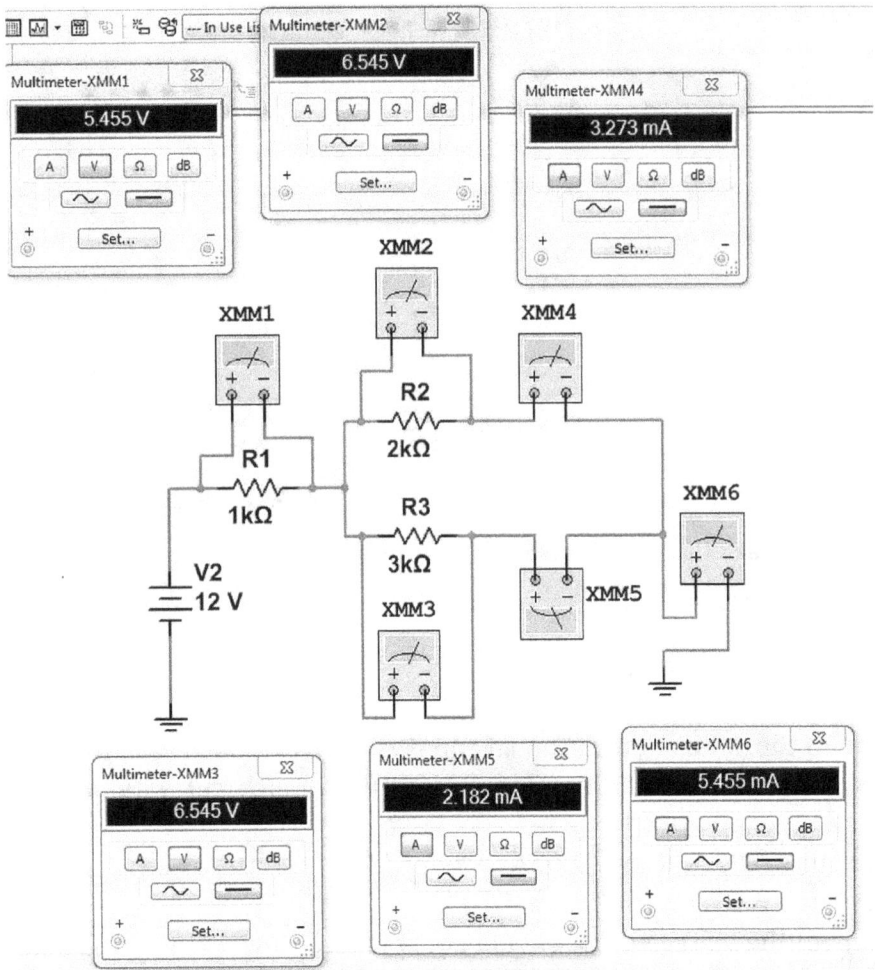

The meters are measuring:
XMM1 = VR1   XMM2 = VR2   XMM3 = VR3
XMM4 = IR2   XMM5 = IR3   XMM6 = IT

Compare the meter measurements with your calculations. Note that there will always be minor rounding errors.

See Appendix E for Series Parallel practice circuits.

Lesson 8-Parallel-Series Circuits

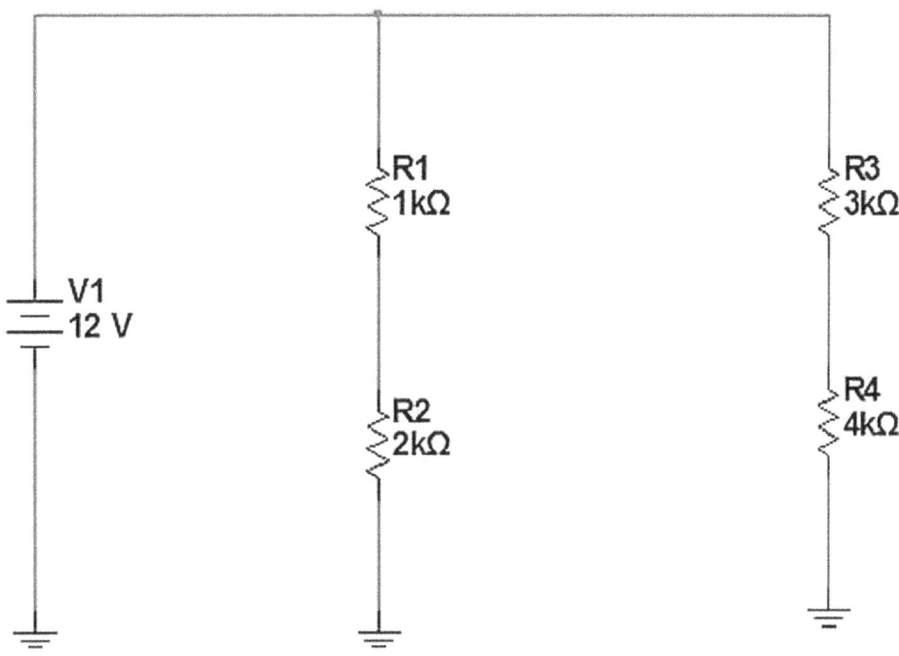

Parallel Series Circuits Points to Remember:

• Each branch is a series circuit that must be solved for each RBX first
• Reduce each series branch to one equivalent resistance by adding the resistances
• The voltage (Vs) on each branch is divided between the series components based on their values
• RT will always be less than the RT of the branch with the least resistance

The recommended procedure to analyze a parallel-series circuit using the above example on the next page:

1. Add the resistance in each branch. R1 + R2=RB1(3KΩ). Then R3 + R4=RB2(7KΩ)

2. The branch total resistances are in parallel so 1/RB1 + 1/RB2 = 1/sum. Invert the sum (1/sum) to get RT. Therefore 1/3kΩ + 1/7KΩ = .0004762. 1/.0004762 = 2100Ω which is RT.

3. Determine IT. IT = Vs/RT which equals 12V/2100Ω = 5.71mA

4. Calculate IB1. IB1 = Vs/RB1 which equals 12V/3KΩ = 4mA

5. Calculate IB2. IB2 = Vs/RB2 which equals 12V/7KΩ = 1.71mA

6. Add IB1 + IB2 = IT. This IT and the IT determined in step 3 should be the same.

7. Calculate VR1. VR1 = IB1*R1 which equals 4mA*1kΩ = 4V

8. Calculate VR2. VR2 = IB1*R2 which equals 4mA*2kΩ = 8V

9. Add VR1 + VR2. This sum should equal Vs

10. Calculate VR3. VR3 = IB2*R3 which equals 1.71mA*3kΩ = 5.13V

11. Calculate VR4. VR4 = IB2*R4 which equals 1.71mA*4kΩ = 6.84V

12. Add VR3+VR4. This sum should equal Vs

13. Calculate PT. PT = IT*VT which equals 5.71mA*12V=68.52mW

14. Calculate PB1. PB1 = VB1*IB1 which equals 12V*4mA = 48mW

15. Calculate PB2. PB2 = VB1*IB2 which equals 12V*1.71mA = 20.5mW

Adding the two branch powers should equal PT

See Appendix F for Parallel Series practice circuits.

Lesson 9-Power in Resistive Circuits

Power in purely resistive circuits is measured in Watts and is true working power. This power is a product of the total circuit current and total circuit voltage. Or, the power in a single resistance can be calculated if needed, using the Ohm's Law formulas and two known quantities with the particular resistance.

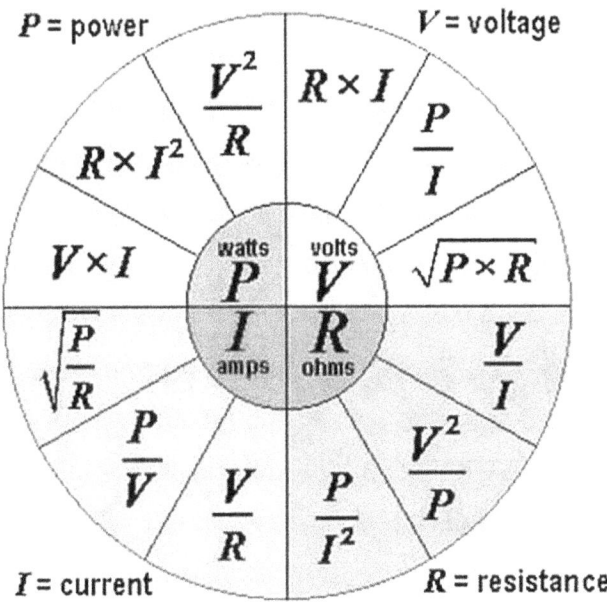

Circuit Power Points to Remember:

- Use the P (watts) section of the Ohm's Law wheel to calculate power
- Power can be calculated for an entire circuit or an individual resistance
- The sum of the individual resistive powers will equal PT
- The largest resistance will dissipate (use) the most power
- Power in purely resistive circuits is True power
- Power is constant in DC circuits with an unchanging RT or Vs
- The more True power that is used in a circuit the more heat that will be generated
- The physical size of a resistor determines how much wattage it can handle
- Wattage rating of a resistance doesn't affect circuit analysis

Lesson 9a-PT with IT & VT Known

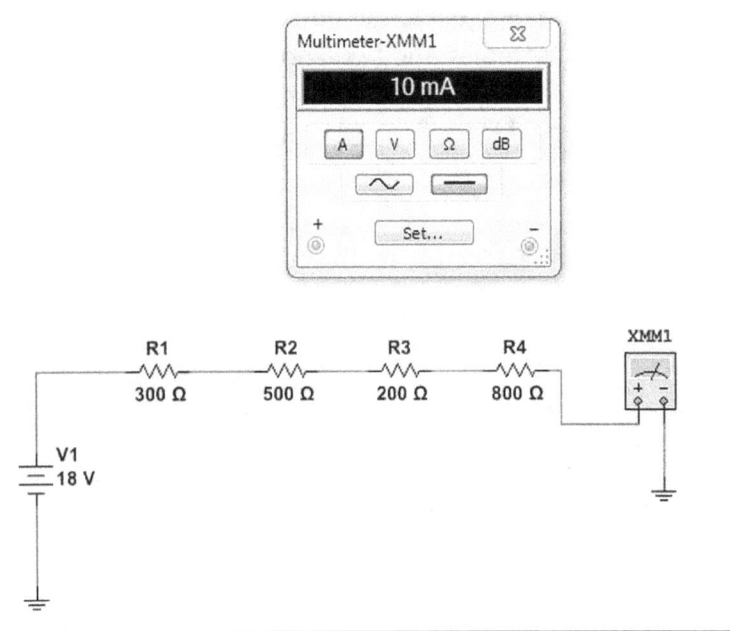

In the above circuit our two known quantities are VT and IT.

Therefore:
PT = VT x IT which equals 12V * 12mA (.012A) = .144W or 144mW

The power in each resistor is:

PR1 = 12mA^2 * 100Ω (R1) = .0144W or 14.4mW

PR2 = 12mA^2 * 200Ω (R2) = .0288W or 28.8mW

PR3 = 12mA^2 * 300Ω (R3) = .0432W or 43.2mW

PR4 = 12mA^2 * 400Ω (R4) = .0576W or 57.6mW

When the power in each resistor is added the sum should equal the previously calculated PT.
PR1 + PR2 + PR3 + PR4 = 144mW

Lesson 9b-PT & PRX with RT & VT Known

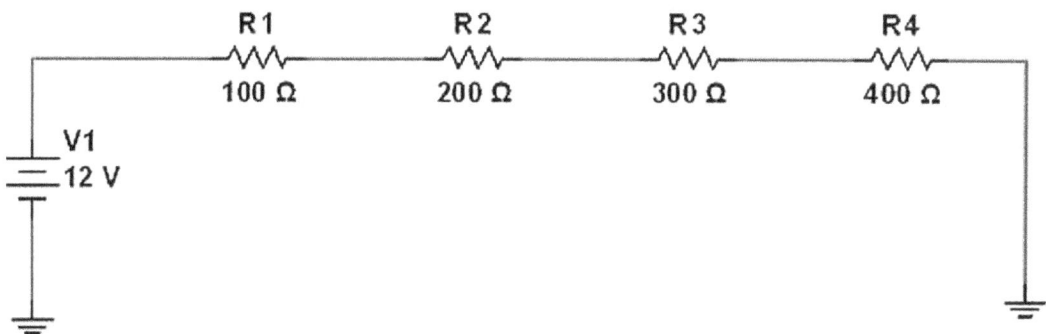

Calculate RT. RT = R1 + R2 + R3 +R4 = 1000Ω (1KΩ)

In the above circuit our two known quantities are VT and RT. Therefore:
PT = V_s^2/RT = 144mW

For each resistor power:

PRx = (Rx/RT) * PT

PR1 = (100Ω/1KΩ) * 144mW = 14.4mW

PR2 = (200Ω/1KΩ) * 144mW = 28.8mW

PR3 = (300Ω/1KΩ) * 144mW = 43.2mW

PR4 = (400Ω/1KΩ) * 144mW = 57.6mW

Lesson 9c-PT & PRX with RT & IT Known

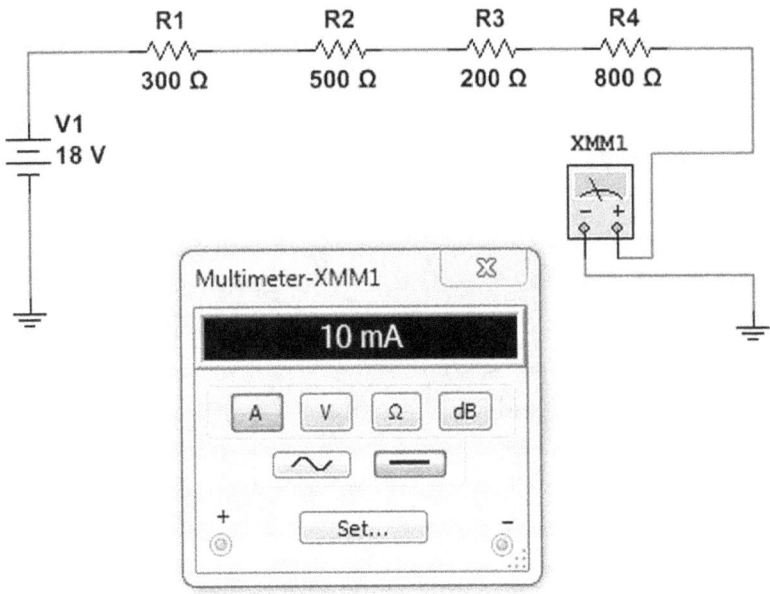

RT = R1 + R2 + R3 + R4 = 1800Ω (1.8KΩ)

PT = IT2*RT = 10mA*1.8KΩ = .180W or 180mW

To completely analyze the circuit:

VT = PT/IT = 180mW/10mA = 18V

VR1 = IT*R1 which equals 10mA*300Ω = 3V

VR2 = IT*R2 which equals 10mA*500Ω = 5V

VR3 = IT*R3 which equals 10mA*200Ω = 2V

VR4 = IT*R4 which equals 10mA*800Ω = 8V

VT = VR1 + VR2 + VR3 + VR4 = 18V (Used to check the math for the voltage drops)

PR1 = IT*VR1 which equals 10mA*3V = 30mW

PR2 = IT*VR2 which equals 10mA*5V = 50mW

PR3 = IT*VR3 which equals 10mA*2V = 20mW

PR4 = IT*VR4 which equals 10mA*8V = 80mW

PT = PR1 + PR2 + PR3 + PR4 = 180mW (Used to check the power calculations)

Lesson 9d-PT in Parallel Circuits

Parallel Circuits Power Points to Remember:

- All branches have the same voltage
- PT is IT x VT
- PT is also the sum of the branch wattages

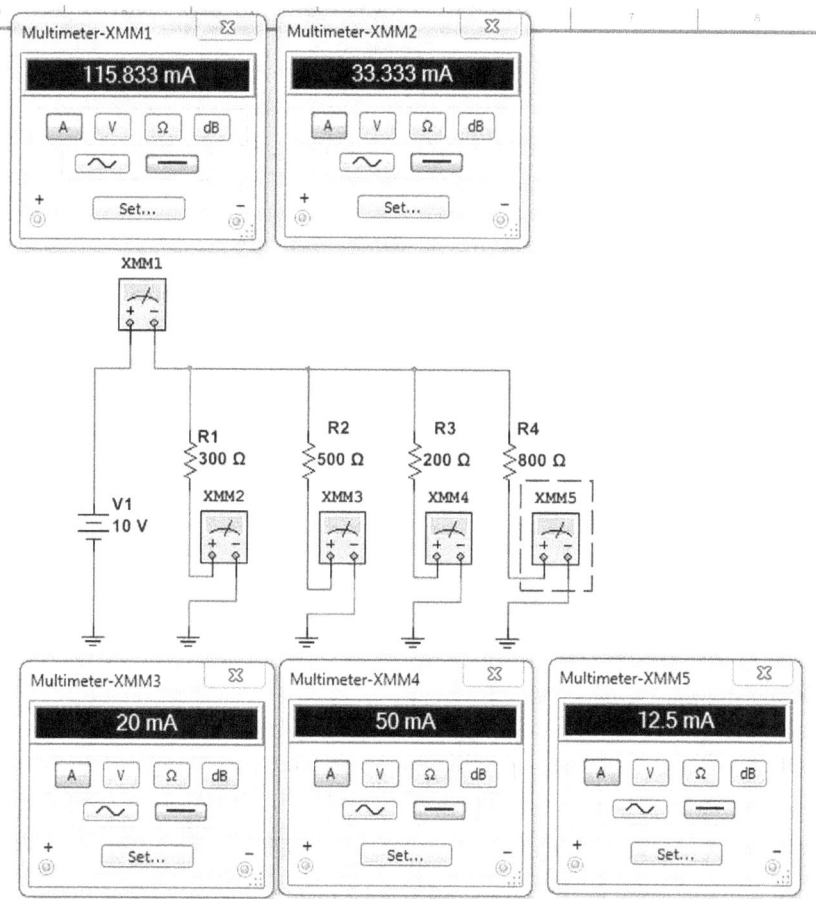

In this circuit Amp meters have been placed to measure the total current and the current in each branch.

The meters are measuring:
XMM1 is IT, XMM2 is IR1, XMM3 is IR2, XMM4 is IR3, XMM5 is IR4

PT = 115.8mA*10V = 1.158W
PR1 = 33.33mA*10V = 333.3mW
PR2 = 20mA*10V = 200mW
PR3 = 50mA*10V = 500mW
PR4 = 12.5mA*10V = 125mW
PT = PR1 + PR2 + PR3 + PR4

Lesson 10-Series Resistive Voltage Divider Circuit

A Voltage Divider is simply a circuit with the voltage "tapped off" at points in the circuit measured from each point to ground. Typically this is a series circuit.

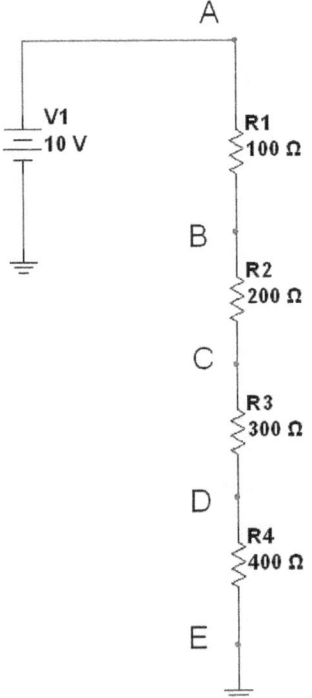

Presented will be a method of analysis of and determining the available voltage at any tap point.

Each tap point is indicated with a letter.
Point A will indicate the source voltage which in this case is 10V.

Point B - The voltage VB will be VA - VR1
Point C - The voltage VC will be VB - VR2
Point D - The voltage VD will be VC - VR3
Point E - The voltage VE will be VD - VR4. This value should be 0V or close to 0V with rounding errors.

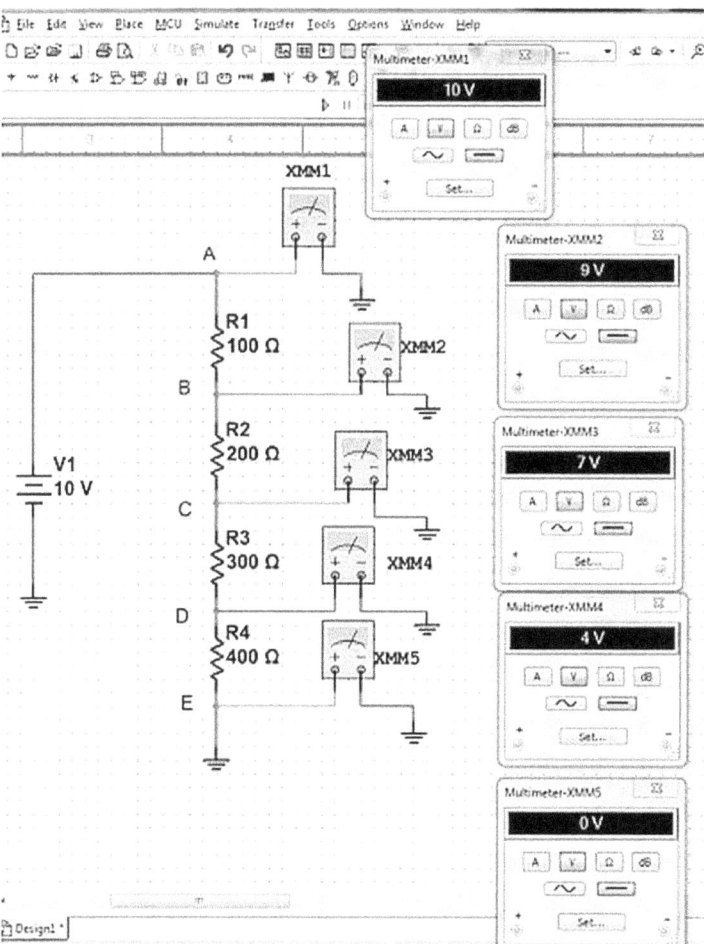

The complete circuit analysis:

RT = 100Ω + 200Ω + 300Ω + 400Ω = 1kΩ
IT = 10V/1KΩ = 10mA
VR1 = 10mA*100Ω = 1V
VR2 = 10mA*200Ω = 2V
VR3 = 10mA*300Ω = 3V
VR4 = 10mA*400Ω = 4V
Therefore:
VB = VA-VR1 = 9V

VC = VB-VR2 = 7V
VD = VC-VR3 = 4V
VE = VD-VR4 = 0V

Remember that an actual circuit with these values may not measure exactly the values indicated but they should be very close. Resistors rarely measure their exact band value.

Lesson 11-Series and Parallel DC Voltage Sources, Amp Hour Ratings

Voltage Sources Points to Remember:

- DC sources (typically batteries) add in series
- The Ah rating of each battery do not add in series
- DC sources in parallel do not add the individual voltages
- The Ahr rating of each battery adds in parallel
- When doing circuit analysis reduce the voltages sources to one source for simplicity

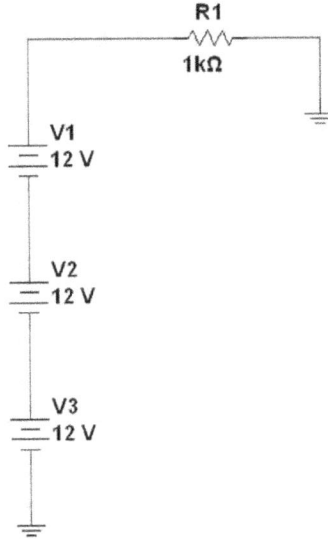

To get VT with series voltage sources simply add them.
V1 + V2 + V3 = 36V

Lesson 11a Amp Hour Ratings

The amp hour rating of a battery is how long in hours (or minutes) a battery can supply current to a circuit. In series sources the Ahr ratings do not add so in the circuit above if each battery had a 1Ahr rating then that is what is used in the calculation, not 3Ahrs. To calculate how long the above circuit will run:

IT = VT/RT which equals 36V/1KΩ = .036A or 36mA

Runtime = Ahr rating/Circuit current = 1Ahr/.036A = 27.78 hrs

With voltage sources in parallel the Ahr ratings add. The VT is the value of any one DC source. So if each battery is 12V and has a 1Ahr rating then the VT is 12V and the total Ahr rating is 3Ahrs.

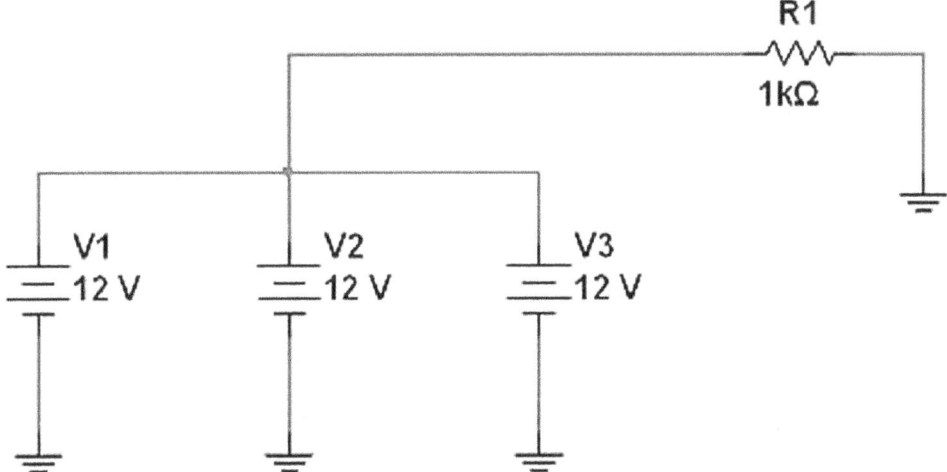

How long will the batteries run the circuit?

IT = 12V/1KΩ = .012A or 12mA

Runtime = 3Ahr/.012A = 250 hrs

Lesson 12 Parallel Current Sources

With these type of circuit we will solve for current in the load resistor (R3 in this instance) by calculating current from each DC sources. Resistor R3 will see the current from both sources.

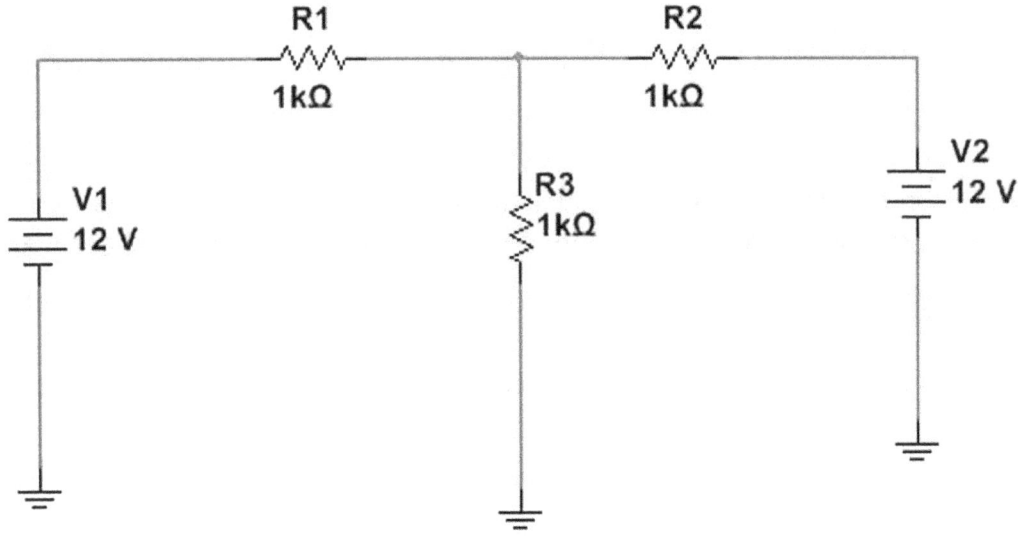

The first step in analysis of this circuit configuration is to replace one of the current sources with a short circuit, turning the circuit into a series parallel circuit. Here is the analysis procedure: (Circuit analysis is on the next 2 pages)

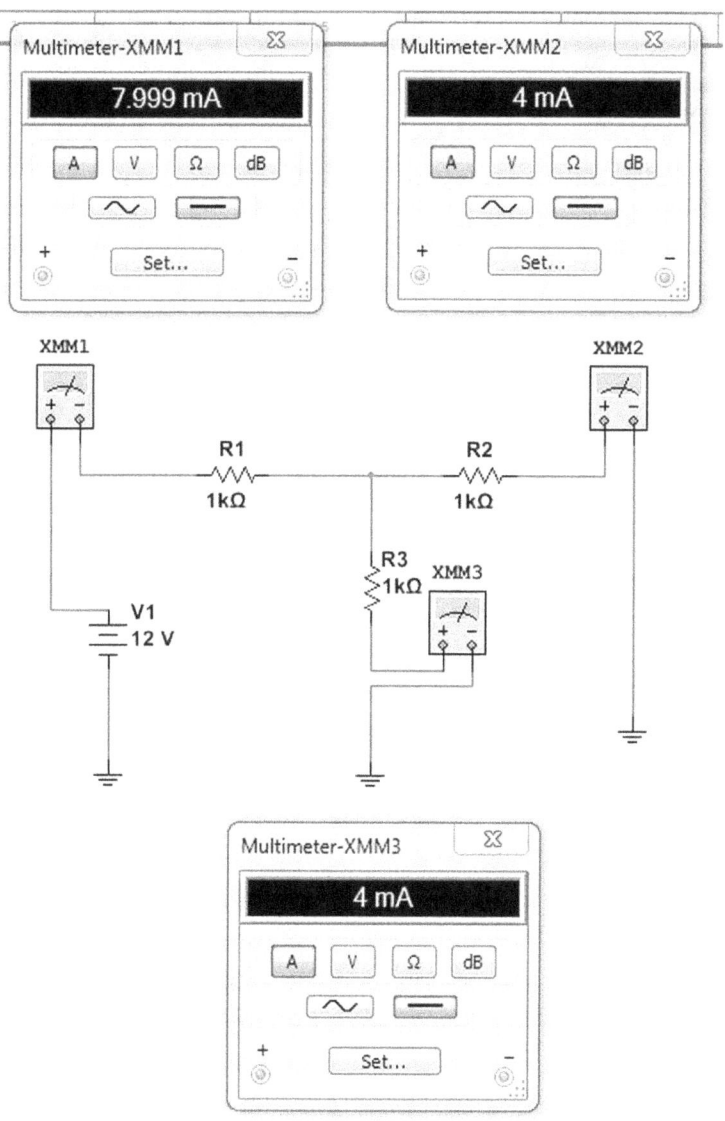

1. Replace one of the current sources with a short circuit.
2. Solve the parallel branch for R parallel (1/R2+1/R3=500Ω)
3. Add R1 to R parallel to get RT. (1KΩ+500Ω=1500Ω)
4. IT=Vs/RT which equals 12V/1500Ω = 8mA
5. Calculate V parallel. V parallel = 8mA*500Ω=4V
6. Calculate IR3. IR3 = 4V/1KΩ=4mA

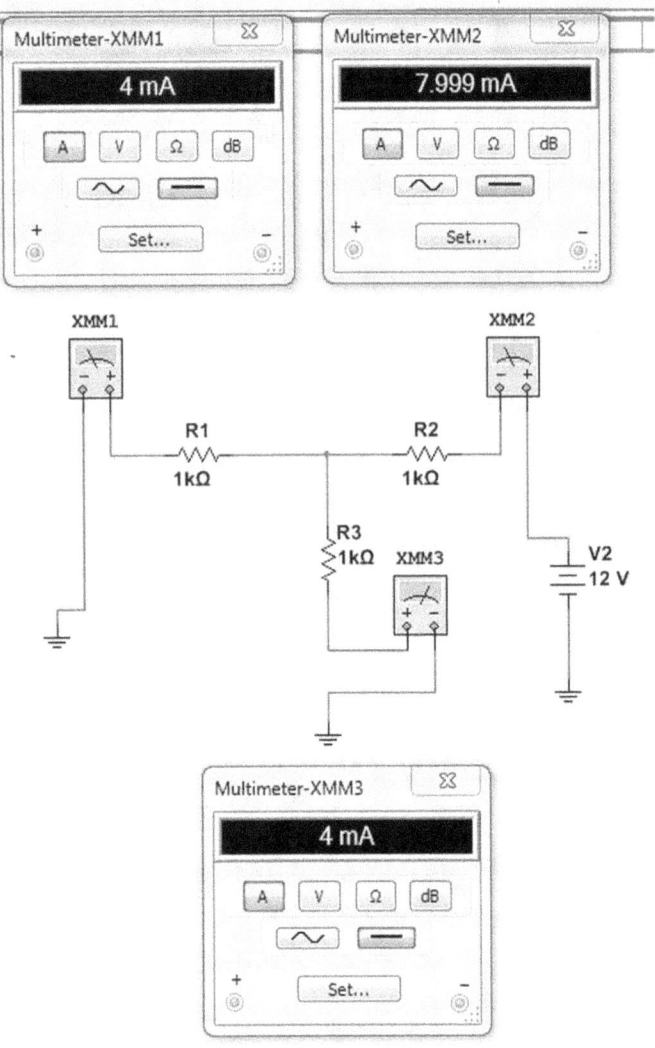

7. Now restore the current source that was shorted and short the other current source.
8. Solve the parallel branch for R parallel (1/R1 + 1/R3 = 500Ω)
9. Add R2 to R parallel to get RT. (1KΩ + 500Ω = 1500Ω)
10. IT = Vs/RT = 8mA
11. Calculate V parallel. V parallel = 8mA*500Ω = 4V
12. Calculate IR3. IR3 = 4V/1KΩ = 4mA

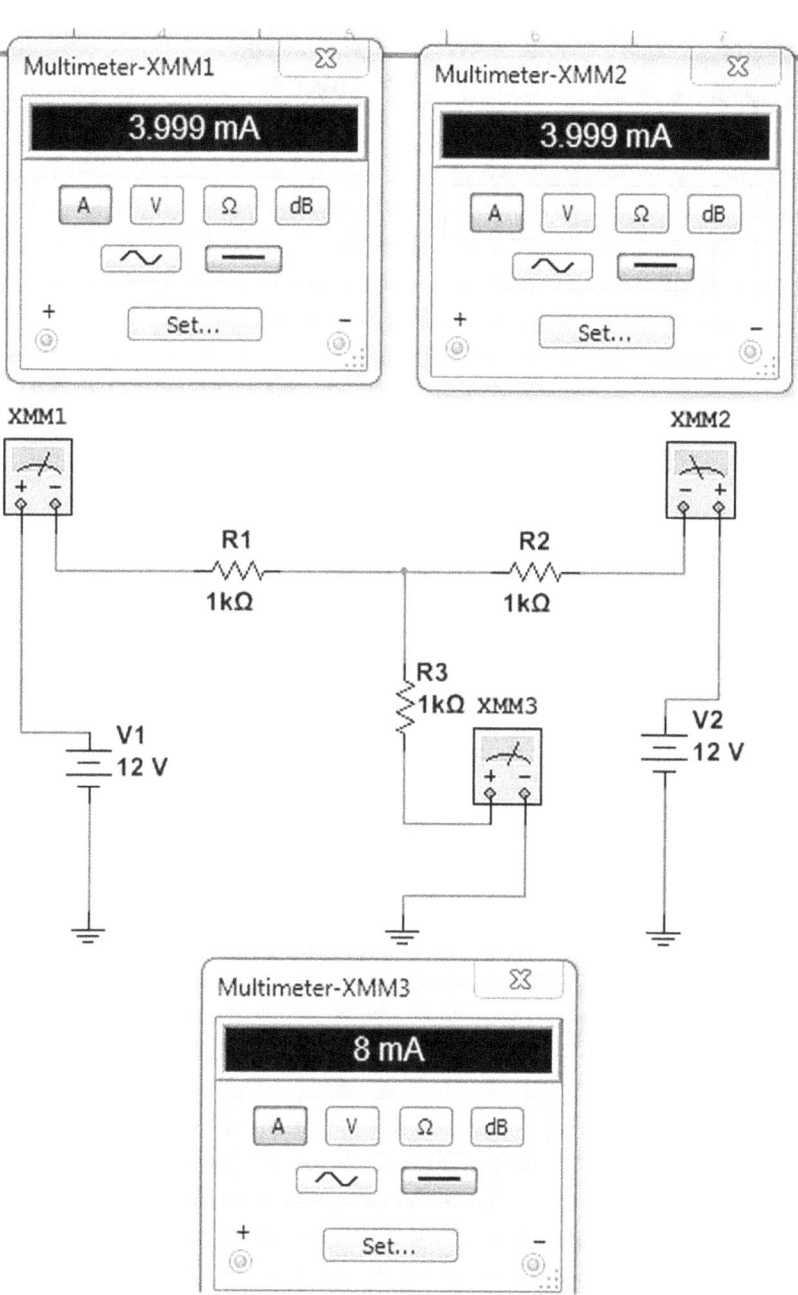

The complete current measurements for the study circuit above.

The end result is that R3 sees the sum of the currents through R1 and R2.

Appendix A Series Circuit Analysis Unknown Quantities Flowchart

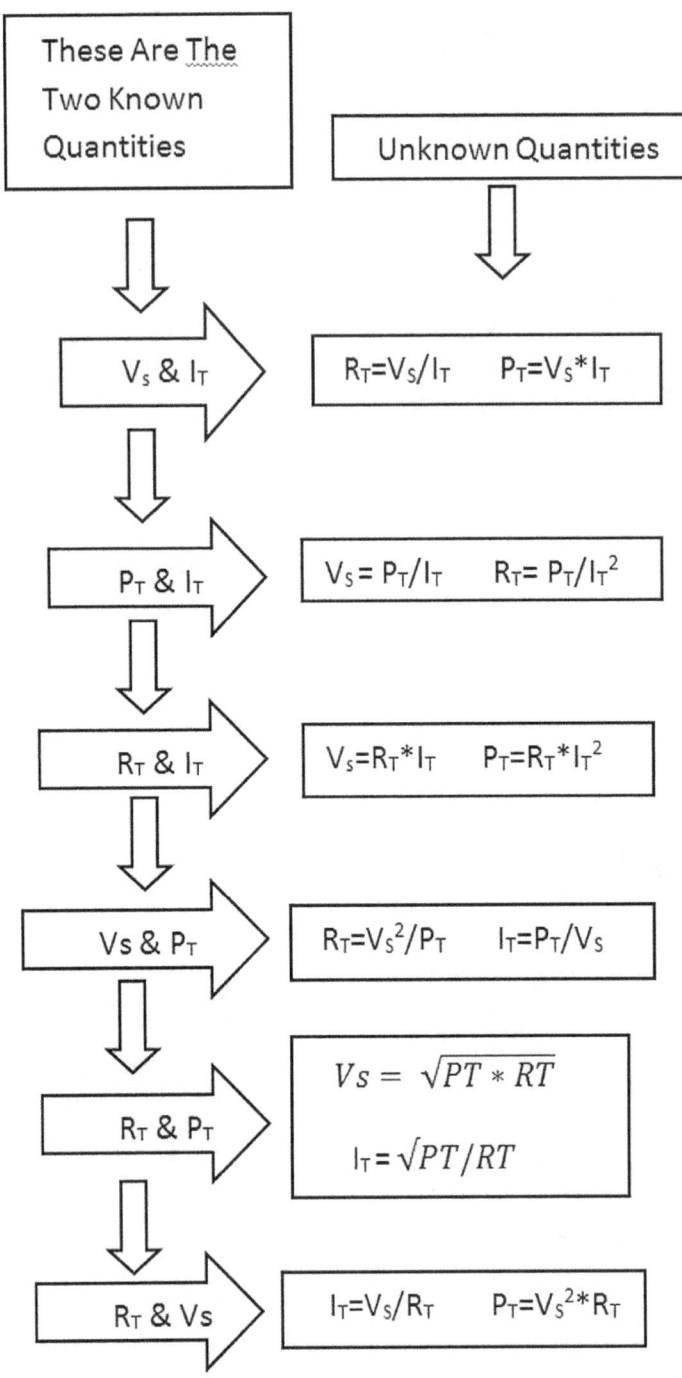

1st Semester Electronics

Appendix B Practice Series Circuits

Practice Series Circuits. Answers are given at the end of the exercise.

Basic formulas that will be used:

I=V/R   V=I*R   PT=V*1   PT=PR1+PR2...............

Circuit 1

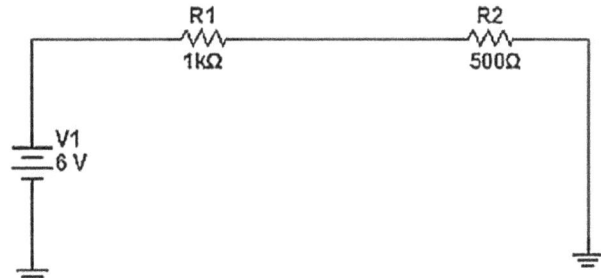

RT=_____   IT=_____   VR1=_____

VR2=_____   PR1=_____   PR2=_____

PT (PR1 ÷ PR2)=_____   PT (Vs*IT)=_____

- Notice that the larger value resistor will have the largest voltage drop.

Circuit 2

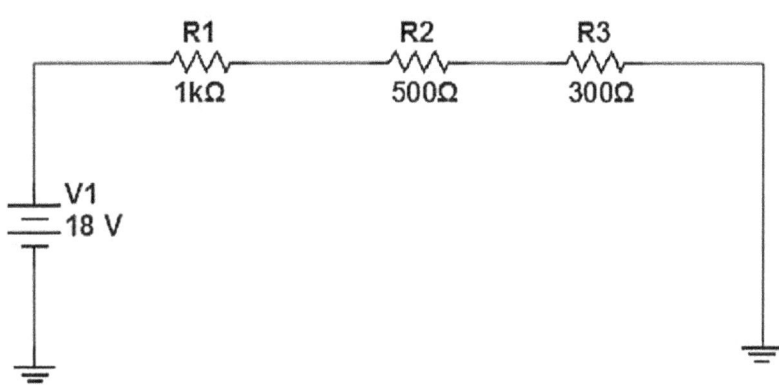

RT=   IT=   VR1=   VR2=   VR3=

PR1=   PR2=   PR3=   PT (PR1 + PR2+PR3)=

Circuit 3

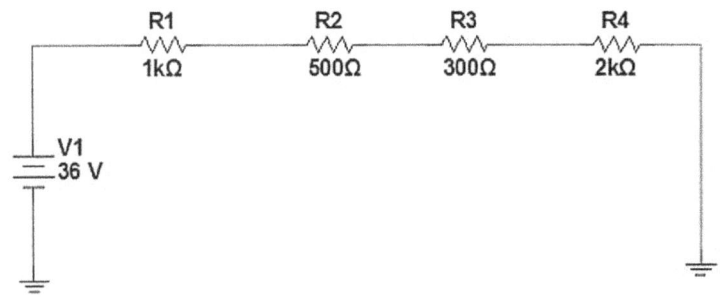

RT=   IT=   VR1=

VR2=   VR3=   VR4=

PR1=   PR2=   PR3=

PR4=   PT =

Series Circuits Answers

Circuit 1

RT = 1.5KΩ   IT = 12mA   VR1 = 12V   VR2 = 6V

PR1 = 144mW PR2 = 72mW PT = 144mW

Circuit 2

RT = 1.8kΩ   IT = 10mA   VR1 = 10V   VR2 = 5V

VR3 = 3V   PR1 = 100mW PR2 = 50mW PR3 = 30mW

PT = 180mW

Circuit 3

RT = 3.7KΩ   IT = 9.47mA   VR1 = 9.47V   VR2 = 4.73V

VR3 = 2.84V   VR4 = 18.97V   PR1 = 89.7mW PR2 = 44.8mW

PR3 = 26.9mW

Appendix C Parallel Practice Circuits

Circuit 1

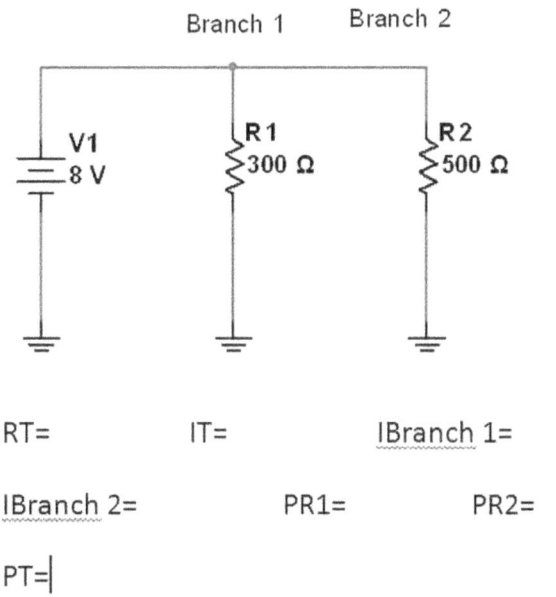

RT=   IT=   IBranch 1=

IBranch 2=   PR1=   PR2=

PT=

Circuit 2

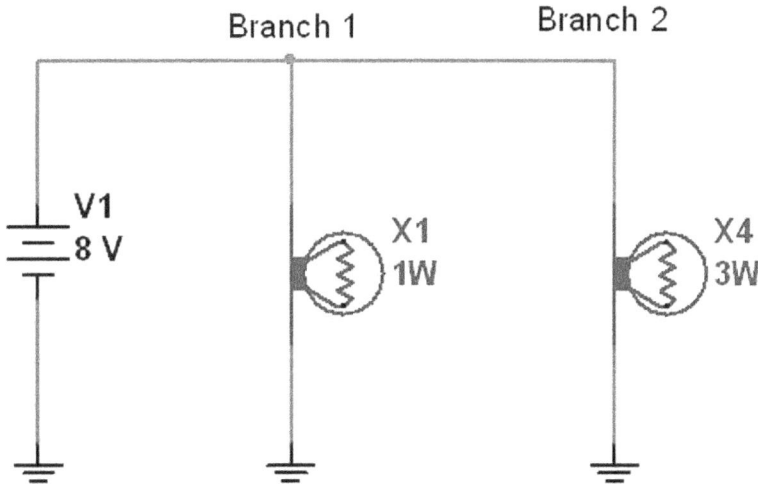

I Branch1 =   I Branch2 =   IT =

RX1 =   RX4 =   RT =

Circuit 3

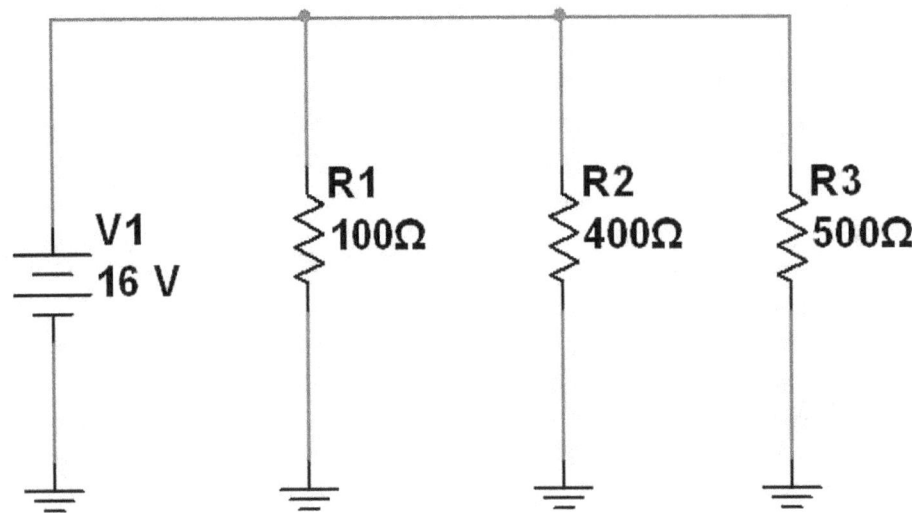

R_T =   I_T =   I_R1 =

I_R2 =   I_R3 =   P_R1 =

P_R2 =   P_R3 =   P_T =

Answers to Parallel Practice Circuits

Circuit 1

R_T = 187.5Ω   it = 42.7mA   I branch1 = 26.7mA

I Branch2 = 16mA P_{R1} = 213.6mW P_{R2} = 128mW

P_T = 341.6mW

Circuit 2

I Branch1 = 125mA   I Branch2 = 375mA I_T = 500mA

R_{X1} = 64Ω R_{X2} = 21.3Ω R_T = 16Ω P_T = 4W

Circuit 3

R_T = 69Ω I_T = 231.9mA I_{R1} = 160mA

I_{R2} = 40mA I_{R3} = 32mA P_{R1} = 2.56W

P_{R2} = 640mW P_{R3} = 512mW P_T = 3.71W

Appendix D Series Parallel Practice Circuits

Circuit 1

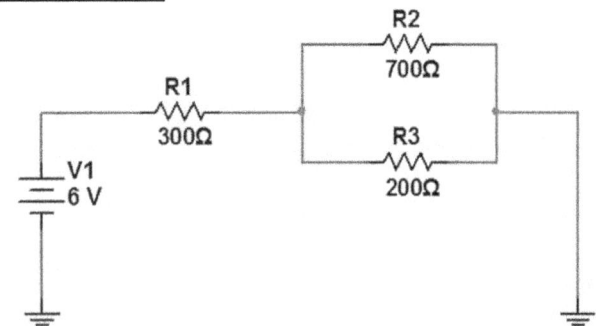

R parallel=   RT=   IT=

VR1=   V parallel= I   IR2=

IR3=   PR1=   PR2=

PR3=   PT=

Circuit 2

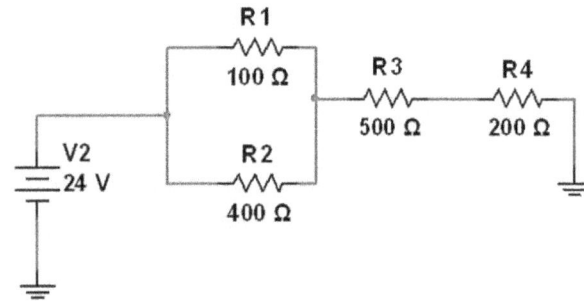

R parallel=   RT=   IT=

V parallel=   VR3=   VR4=

IR1=   IR2=

PR1=   PR2=   PR3=

PR4=   PT=

Practice Series Parallel Circuits Answers

Circuit 1

R parallel= 155.6Ω   RT= 455.6Ω   IT= 13.17mA

VR1= 3.95V   V parallel= 2.05V   IR2= 2.93mA

IR3= 10.3mA   PR1= 52mW   PR2= 6mW

PR3= 21.1mA   PT= 79mW

Circuit 2

R parallel= 80Ω   RT= 780Ω   IT= 30.77mA

V parallel= 2.46V   VR3= 15.4V   VR4= 6.15V

IR1= 24.6mA   IR2= 6.15mA

PR1= 60.52mW   PR2= 15.13mW   PR3= 473.9mW

PR4= 189.24mW   PT= 738.5mW

Appendix E Parallel Series Circuits

Circuit 1

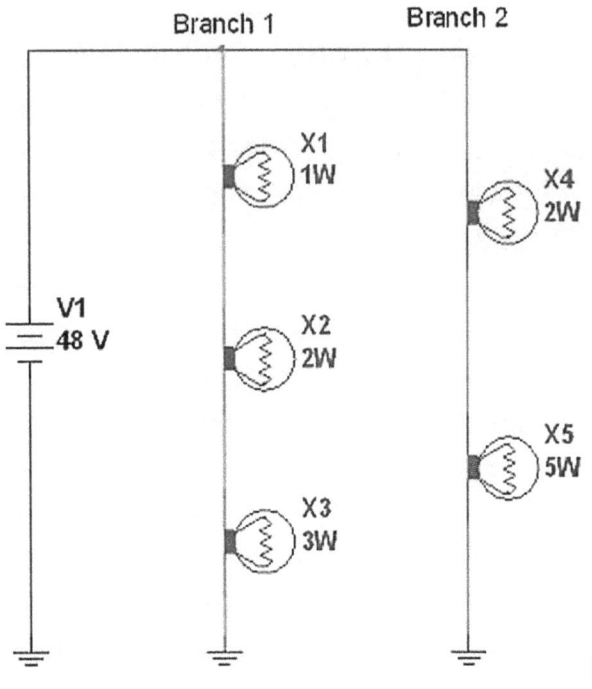

$P_{T \text{ Branch 1}} =$ _____ $P_{T \text{ Branch 2}} =$ _____

$P_T =$ _____ $I_{\text{Branch 1}} =$ _____ $I_{\text{Branch 2}} =$ _____

$V_{x1} =$ _____ $V_{x2} =$ _____ $V_{x3} =$ _____

$V_{x4} =$ _____ $V_{x5} =$ _____

Circuit 2

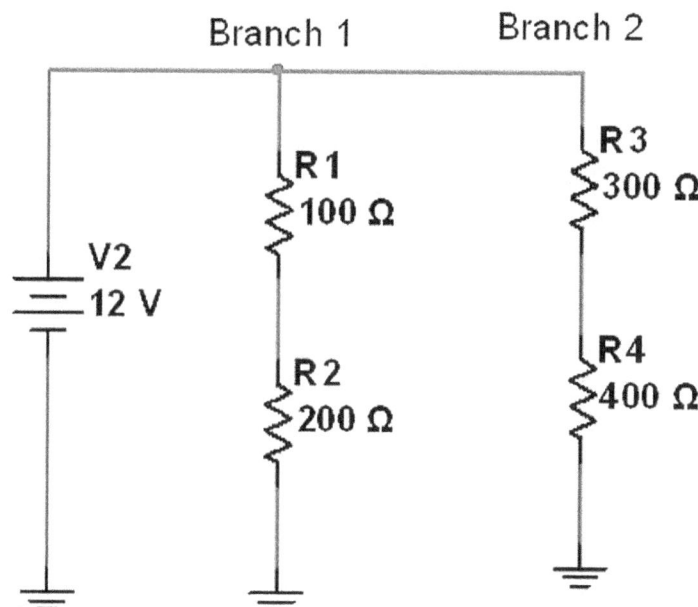

R branch1=   R branch2=

I branch1=   I branch2=

IT=   PT=

VR1=   VR3=

VR2=   VR4=

PR1=   PR3=

PR2=   PR4=

Circuit 3

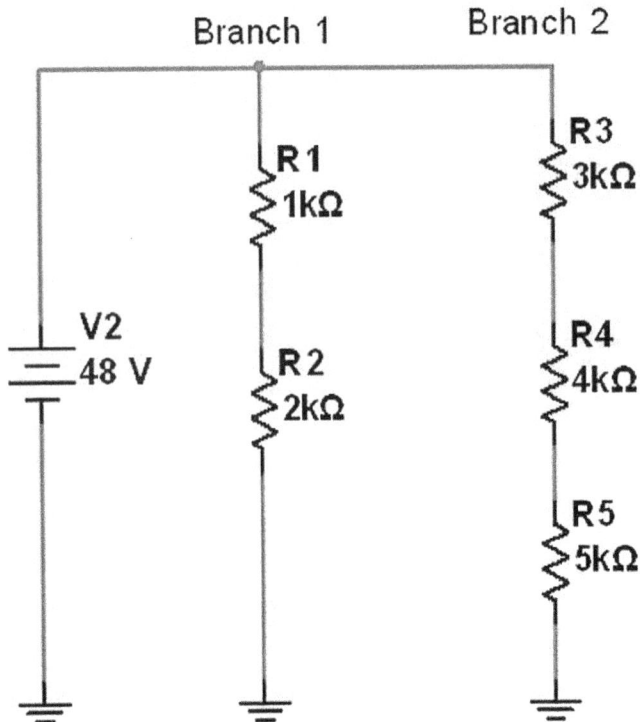

I branch1=   I branch2=

VR1=   VR3=

VR2=   VR4=

IT=   VR5=

P branch1=   P branch2=

PT=

Answers to Parallel Series Circuits

Circuit 1

PT Branch1=6W PT Branch2=7W
PT=13W   I Branch1=125mA
I Branch2=145.8mA IT=207.8mA
Vx1=8V   Vx2=16V   Vx3=24V
Vx4=13.7V   Vx5=34.3V

Circuit 2

R Branch1=300Ω R Branch2=700Ω
RT=210Ω
I Branch1=40mA I Branch2=17.14mA
IT=57.14mA PT=685.7mW
VR1=4V VR3=5.14V
VR2=8V VR4=6.86V
PR1=160mW PR3=88mW
PR2=320mW PR4=117.6mW

Circuit 3

R Branch1=3KΩ R Branch2=12KΩ
I Branch1=16mA I Branch2=4mA
RT=2.4KΩ
VR1=16V VR3=12V
VR2=32V VR4=16V
IT=20mA VR5=20V
P Branch1=768mW P Branch2=192mW
PT=960mW

Appendix F

Practice Conversions

Convert the following:

45ma =._____ Amps
677mV = _____V
5000mV = _____V
20uA = _____ mA
1200 Ohms = _____K Ohms
5560 Ohms = _____K Ohms
4.05K Ohms = _____Ohms
500uA = _____ mA
.450mA = _____uA
.5M Ohms = _____Ohms
6.5A = _____mA
18000V = _____KV
9500uV = _____mV
4500nA = _____uA
.000070V= _____uV
37KV= _____ V
.000125A= _____uA
650pA= _____ A
10pA= _____A
.050V= _____mV
.000009V= _____uV
.0009V= _____mV
.000089A= _____uA
.000000015A= _____pA
.225A = _____uA
.0000005A= _____uA

Answers to Conversions

45ma = .045 Amps
677mV = .677V
5000mV = 5V
20 uA = .02mA
1200 Ohms = 1.2K Ohms
5560 Ohms = 5.56K Ohms
4.05K Ohms = 4050 Ohms
500uA = .5 mA
.450mA = 450 uA
.5M Ohms = 500,000 Ohms
6.5A = 6500mA
18000V = 18KV
9500uV = 9.5mV
4500nA = 4.5uA
.000070V= 70uV
37KV= 37000V
.000125A= 125uA
650pA= .000000000650A
10pA= .000000000010A
.050V= 50mV
.000009V= 9uV
.0009V= .9mV
.000089A= .089mA
.000000015A= 15nA
.225A = 225mA
.0000005A= .5uA

About the Author:

Mark Gray has been a lifelong resident of southeastern North Carolina, having been born in Southport, North Carolina and has resided in Wilmington, North Carolina for the last 14 years.

He is certified by the Electronics Technician's Association of Greencastle, Indiana as a Master Certified Electronics Technician and holds 14 additional technical certifications in areas such as Fiber Optics, Telecommunications, Networking Technologies, Cabling Technologies and Computer Technology.

Mark is the Lead Instructor for the Electronics Engineering Technology program at Cape Fear Community College in Wilmington North Carolina and was previously an adjunct instructor in the Computer Engineering dept.

Mark has held various technical positions at Bellsouth Telecommunications, the New Hanover County North Carolina school system, Corning Inc. and DAK Americas.

Among his other interests are Tae Kwon Do (2nd degree black belt and instructor) scale modelling, genealogy and metal detecting. Mark is also a member of American Mensa and is a testing proctor for that organization.

www.ingramcontent.com/pod-product-compliance
Lightning Source LLC
Chambersburg PA
CBHW081856170526
45167CB00007B/3031